THE *Craft* OF *Research*

D0036956

Chicago Guides
to *Writing*, Editing,
and Publishing

THE *Craft* OF *Research*

WAYNE C. BOOTH

GREGORY G. COLOMB

JOSEPH M. WILLIAMS

THE UNIVERSITY OF CHICAGO PRESS

Chicago & London

The University of Chicago Press, Chicago 60637
The University of Chicago Press, Ltd., London

©1995 by The University of Chicago
All rights reserved. Published 1995
Printed in the United States of America

04   03   02   01   00   5
ISBN: 0-226-06583-9 (cloth)
ISBN: 0-226-06584-7 (paper)

Library of Congress Cataloging-in-Publication Data

Booth, Wayne C.
    The craft of research / Wayne C. Booth, Gregory G. Colomb, Joseph M.
Williams.
        p.   cm. — (Chicago guides to writing, editing, and publishing)
    Includes bibliographical references and index.
    1. Research—Methodology.   2. Technical writing.   I. Colomb, Gregory G.
II. Williams, Joseph M.   III. Title.   IV. Series.
Q180.55.M4B66   1995
001.4'2—dc20   95-2295

Contents

Preface

WE INTEND THIS BOOK for student researchers, from the newest beginners to graduate and professional students. We hope to

- introduce beginning researchers to the nature, uses, and objectives of research and its reporting;
- guide beginning and intermediate researchers through the complexities of planning, organizing, and drafting a report that poses a significant problem and offers a convincing solution;
- show all researchers, from beginning to advanced, how to read their reports as their readers will, how to diagnose passages that readers are likely to find difficult, and how to revise them quickly and efficiently.

Other handbooks on research address some of these matters, but this one differs in several ways.

Many current guides recognize that researchers do not move sequentially from finding a topic to stating a thesis to filling in note cards to drafting and revision. As anyone who has done it knows, real research loops back and forth, moving forward a step or two, going back while at the same time anticipating stages not yet begun, then moving forward again. But so far as we know, no guide has tried to show how each part of the process influences all the others—how asking questions about a topic can prepare the researcher to draft, how the process of drafting can reveal problems with an argument, how the elements of a good introduction can send the researcher back to the library for more work.

This book explains how researchers must work at different stages of their project simultaneously, how that overlap can help them understand the problem better, and how they can manage the complexity this process entails. That means, of course, that you must read this book twice, because we will describe not only how earlier stages anticipate later ones, but how later stages motivate earlier ones.

Because research is so complex, we have been explicit about as many steps as possible, including some usually treated as part of a mysterious creative process. Among the issues that we explicitly "unpack" are these:

- how to turn an interest into a topic, that topic into a few good questions, and the answers to those questions into the solution of a problem;
- how to construct an argument that respects readers' desire to know why they should accept your claim;
- how to anticipate the objections of reasonable but skeptical readers and how to qualify your arguments appropriately;
- how to create an introduction that "sells" the significance of your research problem to readers;
- how to write conclusions that leave the reader with a sense not only of your major claim but of its wider significance;
- how to read your own text as others will and thereby know better where and how to revise it.

We know that some beginning researchers may follow our suggestions in ways that may seem mechanical. We are not deeply troubled by that because we believe that it is better for students to succeed mechanically than for them not to succeed at all. We also believe that teachers can trust students to overcome that early and inevitable awkwardness. All of us tend to be mechanical when we first learn a skill, but eventually we learn to hide its machinery behind its substance.

Another distinguishing feature of this book is that we relentlessly encourage researchers to side with their readers, and we explain explicitly how to do that by explaining how readers read. The aim of a research report is to engage in a conversation with those who may not be eager to change their minds, but who, for good reasons, will. The place where you conduct that conversation is in your report. As they read, readers expect to find certain features of organization; they prefer certain patterns of style; they tacitly pose questions, raise objections, usually want to see matters laid out more explicitly than you may think necessary. We believe that if you can understand how readers read and can know better how to meet their expectations, then you have a better chance of helping them to see things your way.

We focus on the process of doing all this by showing how the formal properties of the "product"—the report—can help you plan and conduct the process that creates it. As you will see, the elements of a report, its structure, style, and formal conventions, are not empty formulas that writers mimic just because thousands used those formulas before them. Those forms and patterns are the means by which researchers, beginners and experienced, test their work, explore their understanding of it, even find new directions. In other words, we believe that the formal demands of the product can not only guide the researcher through the process of its creation, but can themselves contribute to the author's creativity.

We have also tried to indicate what researchers at different stages of their professional lives should know and be able to do. If you are doing your first research project, you should have an idea of what advanced researchers are expected to do, but not worry if you cannot do it all. You should know, however, what your teachers are likely to expect of *you,* particularly if you are preparing to become a serious researcher. So we occasionally announce that we are about to address an issue that is particularly important to ad-vanced researchers. Those of you just getting started may be tempted to skip those sections. We hope you won't.

This book was born in the belief that the skills of doing and reporting research are not just learned but can also be taught. Where we could explain the steps in the process explicitly, we did; where we couldn't, we tried to describe its general shape. Some aspects of research can be learned only in the context of a community of researchers committed to particular topics and ways of thinking and interested in sharing the fruits of their work. But when such a context is not available, students can still learn important research skills through direct instruction and carry those skills into the communities they want to join. We explore some specific ways to do that in our Postscript to Teachers.

This book was also born in our experience that research is not the sort of thing one learns once and for all. Each of the three of us has faced research projects that have forced us to take a fresh look at how we do research, even after we had decades of experience. In those moments when we had to adapt to a new research community or to changes in our own, we have used the principles presented here to help us focus on what is most important to readers. So we have written a book that you can return to as you and your circumstances change, one that we hope will be useful time and again as you grow as a researcher.

We want to thank those who have helped us bring this project to fruition. They include some early readers: Steve Biegel, Jane Andrew, and Donald Freeman. The chapter on the visual presentation of data was improved significantly by the comments of Joe Harmon and Mark Monmonier. We would also like to thank those who helped us select and edit the Appendix on Sources: Jane Block, Diane Carothers, Tina Chrzastowski, James Donato, Bill McClellan, Nancy O'Brien, Kristine Fowler, Clara Lopez, Kim Steele, David Stern, Ellen Sutton, and Leslie Troutman. We are also indebted to those at the University of Chicago Press who, when we agreed to undertake this project almost a decade ago, kept after us until we finally delivered.

From WCB: In addition to the hundreds who have taught me whatever I have contributed to this book, I should like to thank my wife, Phyllis, my two daughters, Katherine and Alison, my three grandchildren, Emily, Robin, and Aaron: together these six keep me optimistic about the future of responsible inquiry.

From GGC: Through turbulent times and calm, through creative periods and fallow, there was always home and family—Sandra, Robin, Karen, and Lauren—to give point and purpose to it all.

From JMW: Joan, Megan, Ol, Chris, Dave, and Joe have sustained me when we have been together and when not. Together is better.

PART ONE

Research, Researchers, and Readers

Prologue: Starting a Research Project

IF YOU ARE BEGINNING your first research project, the task probably seems overwhelming: How do you look for a topic? Where do you find relevant information? How do you organize it once you find it? And even if you have already written a research paper in a writing class, the idea of another one may be still more intimidating if you now have to do the *real* thing for the first time. Even experienced researchers feel a bit anxious when they undertake a new project, especially if it's of a new kind. So whatever concern you feel now, all researchers have felt, and many of us still do. The difference is that experienced researchers know what lies ahead— hard work, but also the pleasure of the chase; some frustration, but more satisfaction; periods of confusion, but confidence that, in the end, it will all come together.

MAKING PLANS

Experienced researchers also know that, like any complex project, research is more likely to "come together" if they have a plan, no matter how rough. Before they start, they may not know precisely what they are looking for, but they know in general the kinds of material they will need, how to find them, and how to use them. And once they assemble their materials, competent researchers don't just start writing, any more than competent builders just start sawing wood. *They plan a product of a certain kind and certain shape, a product that expresses their deliberate intention to achieve a particular end, a product all of whose parts are designed to contribute to that end.* But neither do good researchers let their plan box them in. They are ready to change their plan if they run into a problem or if they suddenly understand their project better, or if they discover in some by-way a more interesting objective that requires them to head off in a new direction. But they all begin with a purpose and plan of some kind.

In fact, writing projects of almost every kind begin with a plan to create a document of a specific form, often a form shaped by the experience of generations of writers. Writers use these forms not just to please editors or supervisors, but to save themselves from having to invent a new form for each new project and, just as important, to help readers recognize their objectives. A reporter knows that she has to use an inverted pyramid form in a news story, putting the most salient information first, not for *her* benefit but so that *we* can find the gist of the news quickly and then decide whether to read on. The form of an audit report tells an accountant what he has to include, but it also helps *investors* find what they need in order to evaluate Abco, Inc. as an investment; the nurse knows what to write in a patient record so that *other* caregivers can use it; a police officer composes her arrest report in a standard way so that it can guide those who will investigate the crime later. In the same way, when a researcher reports her results in a form familiar to readers, those readers can read her report most efficiently.

Within these forms, of course, writers are free to take different points of view, emphasize different ideas, and put a personal stamp on their work. But when they follow a standard plan, they make it easier for themselves to write and for their readers to read.

The object of this book is to help you create and execute such a plan.

THE VALUE OF RESEARCH

But first a question: Aside from a grade, what's in it for you? An answer that some might think idealistic is that research offers the pleasure of solving a puzzle, the satisfaction of discovering something new, something no one else knows, ultimately contributing to the wealth of human knowledge. For the beginning researcher, though, there are more immediate practical benefits. Right now, doing research will help you understand the material you are studying in a way that no other kind of work can match. More distantly, the skills of research and writing that you learn now will enable you to work on your own later—to gather information, organize it into coherent form, and then report it reliably and persuasively, skills indispensable in a time aptly named the "Age of Information."

In any field, you will need the skills that only research can help you master, whether you expect to design the production line or to run it.

The skills of research and writing are no less important to those who use the research of others. These days, that includes just about all of us. We are inundated with information, most of it packaged to suit someone else's commercial or political self-interest. More than ever, society needs people with critical minds, people who can look at research, ask their own questions, and find their own answers. Only when you have experienced the uncertain and often messy process of doing your own research can you intelligently evaluate the research of others. Writing your own paper will help you understand the kind of work that lies behind what experts say and what you find in your textbooks. It lets you experience firsthand how knowledge develops from answers to research questions: how that new knowledge depends on which questions you ask and which you don't; how those questions depend not just on your interests and goals but on those of your readers; how the standard forms for presenting research shape the kinds of questions you ask, even determine those that you *can* ask.

But we must be candid: writing a research paper is demanding. It consists of many tasks, all competing for your attention, often at the same time. However carefully you plan your research, it will follow a crooked path, taking unexpected twists, even looping back on itself. Each stage overlaps with others: all of us draft before we finish our research, continue our research after we begin drafting. Some of us do our best work late in the game, recognizing the problem we have been trying to solve only after we have found its solution. Others move to the drafting stage late, doing more of the trial-and-error work not on paper but in their heads. Each writer is different, and because each project is different as well, no single plan can solve all problems.

As complex as that process is, though, we will work through it step-by-step, so that you can move forward confidently, even as you confront the inevitable difficulties and confusions that every researcher experiences but eventually learns to manage. When you can manage the parts, you can manage the whole, and look forward to more research with greater confidence.

How to Use This Book

The best way to deal with this complexity (and the anxiety it may arouse) is to read through this book once quickly to see what lies ahead. Then, depending on your level of experience, decide which parts of the task look easy or hard *for you*. As you begin your work, read more carefully the chapters relevant to the task at hand. If you are new to research, begin at the beginning. If you are in an advanced course but are not yet thoroughly at home in your field, skim Part I, read II, but concentrate on III and IV. If you are an experienced researcher, you will probably find most helpful Chapter 4 of Part II, 9 and 10 of Part III, and all of IV.

In Part I, we address some issues that those undertaking their first project often raise—why readers expect you to write up research in particular ways (Chapter 1) and why you should conceive of your project not as solitary work but as a conversation with those whose work you will read and then with those who will read yours (Chapter 2).

In Part II, we explore the process of framing your project—how to find a topic, narrow it, question it, and justify it (Chapter 3), how to transform those questions into a research problem (Chapter 4), how to find and use bibliographic sources to guide the search for answers (Chapter 5), and how to think through what you find (Chapter 6).

In Part III we discuss the nature of a good research argument. We begin with an overview of what a research argument is (Chapter 7); then explain what counts as a significant claim and reliable evidence in its support (Chapter 8); explore an abstract but crucial element of a research argument called its "warrant" (Chapter 9); and conclude with a description of how every writer must address objections, stipulate limiting conditions, and express conditions of uncertainty (Chapter 10).

In Part IV we lay out the steps in producing the final written report, beginning with the drafting process (Chapter 11). Next we address a matter not usually raised in books of this kind—how to communicate complex information visually, even information that is not quantitative (Chapter 12). The next two chapters concern testing and revising your organization (Chapter 13) and style (Chapter 14). We then explain how to produce an introduction that persuades readers that your report will be worth their time (Chapter 15). Finally, we spend a few pages reflecting on research beyond the techniques of doing it well: the matter of the ethics of research in a society that increasingly depends on its results.

Between the chapters you will find a number of "Quick Tips," brief sections that complement the chapters. Some Quick Tips are checklists for using what you learn in the chapters, some discuss additional considerations for advanced students, several address matters not covered in the chapters, but all of them add something new.

Research is hard work, but like any challenging job well done, both the process and the results bring immense personal satisfaction. But research and its reporting are also social acts that require you to think steadily about how your work relates to your readers, about the responsibility you have not just toward your subject and yourself, but toward them as well, especially when you believe that you have something to say that is important enough to cause readers to change their lives by changing what and how they think.

CHAPTER ONE

Thinking in Print:

The Uses of Research, Public and Private

WHEN YOU STAND in the reading room of a library, you see around you centuries of research, the work of tens of thousands of researchers who have thought hard about countless questions and problems, gathered information, devised answers and solutions, and then shared them with others. Teachers at all educational levels devote their lives to research, governments spend billions on it, businesses even more. Research goes on in laboratories, in libraries, in jungles, on oceans and under them, in caves, and in outer space. Research and its reporting make up a huge industry in the world today. Bigger yet is reporting on the reports. Those who can neither do reliable research nor reliably report the research of others will find themselves on the sidelines of a world that increasingly lives on information.

1.1 WHY DO RESEARCH?

But you already know about research, because you do it every day. Research is simply *gathering the information you need to answer a question and thereby help you solve a problem.*

PROBLEM: After a day of shopping, you realize your wallet is missing.
RESEARCH: You recall where you've been and start phoning lost and
      found departments.

PROBLEM: You need a new head gasket for a '65 Mustang.
RESEARCH: You call auto parts stores to see who has one in stock.

PROBLEM: You need to know where Betty Friedan was born.
RESEARCH: You go to the library to look her up in *Who's Who*.

PROBLEM: You hear of a new species of fish and want to learn more.
RESEARCH: You search the *New York Times* to find a story about it.

But while most of us do such research every day, few of us have to write up what we find, because our research is usually for our purposes alone. Even so, we have to rely on the research of others who did write up their findings because they anticipated that one day we might need them to solve a problem of our own: the telephone company did research to assemble the telephone directory; the auto parts suppliers did research to assemble their catalogues; the author of the *Who's Who* article did research on Betty Friedan; the *Times* reporter researched fish.

In fact, research done by others determines most of what any of us believes. Of your three authors, only Williams has ever set foot in Australia, but Booth and Colomb believe in Australia: they feel that they *know* it is there, because for a lifetime they have read about it in reports they trust, seen it on reliable maps, and heard about it in person from Williams. No one has ever been to Venus, but good sources tell us that it is hot, dry, and mountainous. Whenever we "look something up," we do research by consulting the research of others, but we can trust what we find only if those who did that research did it carefully and reported it accurately.

In fact, without reliable *published* research, we would be prisoners of what we alone see and hear, locked in the opinions of the moment. No doubt most of our everyday opinions are sound (after all, we derive many of them from our own research and experience). But mistaken ideas, even ugly and dangerous ones, flourish because too many people accept what they hear, or want to believe, on no very good evidence, and when they act on those opinions they can lead themselves—and us—into disaster. Only when we know that we can trust the research of others can we free ourselves from those who by controlling our beliefs would control our lives.

If, as is likely, you are reading this book because a teacher wants you to undertake a project of your own, you may feel that you are doing it just for the exercise. That's not a bad reason. But your project also gives you the chance to join the oldest and most esteemed of human conversations—the conversation conducted by Aristotle, Marie Curie, Booker T. Washington, Albert Einstein, Margaret Mead, the great Islamic scholar Averroës, the Indian philosopher Radhakrishnan, St. Augustine, the students of the Talmud—all those who by contributing to human knowledge have freed us from ignorance and misunderstanding. They and countless others once stood where you now stand. Our world today is different because of their research. It is no exaggeration to say that, done well, yours will change the world tomorrow.

1.2 WHY WRITE IT UP?

Some of you, though, might think that our invitation to join this conversation is easy to refuse. When you report your research, you have to meet a host of strange and complicated requirements, and most of you know that your report will be read not by the world, but only by your teacher. *And, besides, my teacher knows all about my topic. If she just told me the answers or pointed me to the right books, I could concentrate on learning what's in them. What do I gain from* writing up *my research, other than proving I can do it?*

1.2.1 Write to Remember

The first reason to write down what you find is just to remember it. Some exceptional people can gather information without recording it. But most of us get lost when we fill our heads with new facts and arguments: we think about what Smith found in light of Wong's position, and compare both to the odd results in Brunelli, especially as they are supported by Boskowitz, but wait a minute, what did Smith say again? Most of us can answer hard questions only with the help of writing—by listing sources, assembling research summaries, keeping lab notes, and so on. What you don't write down, you are likely to forget, or worse, misremember. That's one reason why researchers don't wait until the end of the process before they start writing: they write from the beginning of their project to its end so that they will better understand what they find and retain it longer.

1.2.2 Write to Understand

A second reason we write is to see more clearly the relationships among our ideas. When you arrange and rearrange the results of your research in new ways, you see new connections and contrasts, complications and implications you might otherwise have missed. Even if you could hold in mind everything you found, you would still need help to line up arguments that pull in different directions, plot out complicated relationships, sort out disagreements among experts. *I want to use these claims from Wong to support my argument, but her argument is undercut by these data from Smith. When I compare them, I see that Smith doesn't address this last part of Wong's argument. Wait a minute—if I introduce it with this section from Brunelli, I can highlight the part of Wong's argument that lets me refute Smith more easily.* Writing induces thinking, not just by helping you understand what you are learning, but by helping you see in it larger patterns of meaning and significance.

1.2.3 Write to Gain Perspective

A third reason we write is that once we get our thoughts out of our heads and onto paper, we see them in a clearer light, one that is always brighter and usually less flattering. Most of us— students and professionals—think our ideas are more coherent while in the warmth of our minds than they turn out to be in cold print. You improve your thinking when you encourage it with notes, outlines, summaries, commentary, and other forms of thinking on paper. But you can reflect on those thoughts clearly only when you separate them from the swift flow of your thinking and fix them in coherent written form.

In short, we write so that we can think better, remember more, and see more clearly. And as we shall see, the better we write, the more critically we can read.

1.3 WHY TURN IT INTO A FORMAL PAPER?

Even if you know that writing is an important part of learning, thinking, and understanding, some of you may still wonder why you need to turn your work into a formal essay or research report. Those forms can pose a problem for students who see no reason to join a conversation they did nothing to create. *Why should I adopt language and forms that are not mine? What's wrong with my language, my concerns? Why can't I report my research in my own way?* Some students even find these expectations threatening: they fear that if they are required to think and write like their teachers, they will somehow become like them.

Such concerns are legitimate, because they touch every aspect of your life. It would be a feeble education that did not affect who and what you are. The deeper your education, the more it will change you. (That's why it is so important to choose carefully what you study and with whom.) But it would be a mistake to think that writing a research paper threatens your identity. Learning to do research will change the way you think by giving you more ways of thinking. You will be different for having done research, because you will be freer to choose who you want to be.

Perhaps the most important reason for reporting research in ways that readers expect is that writing for others is more demanding than writing for yourself. By the time you fix your ideas in writing, they are so familiar that you need help to see them for what they are rather than for what you want them to be. Your best help toward that end is to imagine the needs and expectations of your readers. This is why standard forms and plans are more than convenient vessels into which you pour your findings and conclusions. They help you see your ideas in the brighter light of the knowledge and expectations of your readers, not just so that you can test those ideas, but also to help them grow. You invariably understand your own ideas better when you write to make them accessible to others—arranging your findings in ways that help readers see explicitly how you evaluated your evidence, how you related one idea to another, how you anticipated *their* questions and concerns. All researchers recall a moment when writing for readers revealed a flaw, a blunder, a missed opportunity that escaped them in a first draft written mostly for themselves.

Those who want to join communities that depend on research have to demonstrate not only that they can give good answers to hard questions but also that they can report their results in ways that are *useful* to their community, and that means in ways that are apparent, accessible, and most important, *familiar.* Once you learn the standard forms, you will read more thoughtfully the research reports of others, understand better what your community expects of them, and be able to critique those demands more thoughtfully.

Writing a research paper is, finally, just thinking in print. It gives your ideas the attention they deserve. Written out, your ideas are "out there," disentangled from your memories, opinions, and wishes, ready to be explored, expanded, combined, and understood more fully, because you are cooperating with your readers in a joint venture to create new knowledge. In short, thinking in written form can be more careful, more sustained, more complete, more well-rounded, more attuned to those with different views—more thoughtful—than almost any other kind of thinking.

You can, of course, go through the motions, doing just enough to satisfy your teacher. This book may help you get away with that, but you will cheat yourself if you do. If you find a topic that *you* care about, ask a question that *you* want to answer, find a problem that *you* want to solve, then your project can have the fascination of a mystery story, a story whose solution provides satisfactions that surprise even the most experienced researchers.

CHAPTER TWO

Connecting with Your Reader:

(Re)Creating Your Self and Your Audience

MOST OF THE IMPORTANT THINGS WE DO, we do with others. At first glance, we might think that research is different. We imagine a solitary scholar reading in a hushed library or working in a laboratory surrounded only by glassware and computers. But no place is so full of voices as a library or a laboratory, and even when we seem to be most alone, we work toward an end that always involves us in a conversation with others. We engage with others every time we read a book, use a research apparatus, or rely on a statistical formula. Every time we consult a source, we join and, by joining, sustain a conversation that may be decades, even centuries old.

2.1 CONVERSATIONS AMONG RESEARCHERS

Just as you do in your social life, you make judgments about those with whom you converse as a researcher (just as right now, you should be appraising us three): *Garcia seems reliable if a little predictable; Alhambra is likeable but not careful about her evidence; Wallace has good data, but I don't trust his conclusions.*

These judgments, however, are not one-way—you judging your sources—because they have already judged you, in a sense creating a *persona* for you. These next two passages "create" different readers by crediting them with different levels of knowledge and expertise:

1. The regulation of the interaction of the contractile proteins actin and myosin in the thin filament of the sarcomere by means of calcium blockers is now a common means of controlling cardiac spasms.

2. Your most important muscle is your heart, but it cannot do its work when it is seized by muscle spasms. These spasms can now be controlled with drugs known as calcium blockers. Calcium blockers work in small units of muscle fibers called sarcomeres. Each sarcomere has two filaments, one thick and one thin. The thin filament contains two proteins,

actin and myosin. When actin and myosin interact, your
heart contracts. That interaction is controlled by calcium
blockers.

The first sounds like one expert writing to another; the second like
a doctor carefully explaining complex ideas to a patient.

Your writing will reflect judgments you have made about your
readers' knowledge and understanding, but most importantly, what
you want them to recognize as significant in your research. And
your readers will judge you by how accurately you judge them. If
you misjudge how much background they need, if you offer your
findings in a way that does not speak to their interests, you will
lose the credibility that every writer needs to hold up his side of
the conversation.

So even before you take the first step toward a research paper,
you must think about the kind of conversation you intend to have
with your readers, about the kind of relationship you want to create
with them, about the kind of relationship that you hope they are
willing and able to have with you. That means knowing not only
who they are and who you are, but who you and they think you
are all *supposed to be.*

You may think the answer is obvious—*I know who I am, and
my reader is my teacher*—but student researchers always work in
complicated circumstances. You will seem different in print from
the way you seem in person. And your teachers will react as readers
differently from the way they react in class. Sorting all this out
means recognizing both (1) the different social roles that writers
and readers create for themselves and for each other and (2) the
common concerns that all readers and all writers share.

2.2 Writers, Readers, and Their Social Roles

Your decisions about yourself and your readers are especially
complicated because classroom research assignments create situations that are obviously artificial. If this is one of your first
projects, you are probably not doing it because you really have a
burning need to ask a question whose answer will change the world.
It is equally unlikely that your teacher has asked you to do research
because he really has a burning need for your answer. You are
probably writing to achieve a less direct goal: to learn about research

by *playing the role* of a researcher and *imagining* the role of your reader.

Role-playing is no trivial part of learning. People can learn a skill in three ways: they read about it or hear it explained, watch others practice it, or practice the skill themselves. The most effective learning combines all three, but the third is crucial: not just reading, listening, and watching, but *doing*. And since research is a social activity, practicing research means practicing a social role.

To that end, your paper must create roles for both you and your teacher. But those roles cannot be those of the classroom, where your teacher asks questions so that you can show that you know the answer, or you ask questions because you don't. In your paper, you have to make yourself into a writer/researcher and to cast your teacher as a reader who wants to know, or should want to know, what you have discovered. In fact, you should imagine yourself and your teacher exchanging roles, you as her teacher, she now as your student.

2.2.1 Creating Your Role

For the duration of your research, imagine yourself as someone with information or a claim important enough to pass on to others who should want to know it. In doing this, you may be expected to play the specific role of a professional in the field. If you are in a biology class, for example, you might be expected to keep complete lab notes (including mistakes and dead ends) and, just as an experienced researcher would, to report your results in a professional form. If your project is a family history, you might be expected to report on the literature about your family's ethnic and socio-economic background, just as a professional historian would. Or you may be asked to play the role of an informed person who is not a professional "insider," but rather who you are, a student writing his first research paper in an introductory class.

Your teacher may even create a detailed scenario:

> Write a history of your family for "Project Diversity," part of a centennial celebration and fund-raising campaign: your history and others will appear in a brochure distributed by the Alumni Association to show the diversity of students on campus.

In that scenario, your readers would not be professional historians but potential students and their parents.

But suppose that you were asked to play the role of a researcher reporting to the director of the state Environmental Protection Agency on toxins in the local lake. In that case, you might have to do some research to discover who the director is and how she might use your report. Is her background in politics or in science, and if in science, what kind? Is the report for her alone or to share with the governor? Does she need the information to decide what to do tomorrow or to justify a decision made yesterday?

In short, the first step in preparing to do research is to understand your role in a "scene." Why have you been asked to write this paper? What does your teacher, course, or program want you to learn from it? Is it to give you a taste of what research is like in order to prepare you to major in a field, even to become a professional in it? Or is it to give students seeking a liberal education an occasion to think hard about a topic of their own choice? If you don't know, ask.

Another issue to anticipate is how the *look* of your paper should play a role in its expected social context. In the biology assignment, should the paper take the form of a lab report, or of a decision memo that recommends action? Or of an executive summary? In the case of the history assignment, you would have fewer forms to choose from, but you should know, for example, whether you could build your history around a first-person narrative of your past and its discovery. Or should it be a formal third-person account? Do not begin your research until you know what choices you have about the form of your paper.

2.2.2 Creating a Role for Your Reader

If you have a role to play, so do your readers, one that you must help create for them. Since your teacher is probably your primary reader, you must cast her as someone who, if given good reason, will care about your research problem and want to know its solution. She may also stipulate a role of her own—someone "in" the field who expects you to write as others in the field do. Or, what would be harder, she might play the role of a general reader who does not have an expert understanding of the field and its ways.

Depending on how she casts herself, your teacher will focus on different aspects of your paper. As an expert reader, she will look for correctly formatted citations to the classic studies on the subject; as a general reader, she will want clear, "plain language" explanations of technical terms. And if you are writing a thesis to be read by a committee, you will have to think about roles in ways more complicated yet.

If you are an experienced researcher, you understand how readers differ, but if you are writing your first research paper, you should know that readers adopt roles based on the different ways they will *use* your research. The most important differences are among those who read for diversion, those who want a solution to a practical problem, and those who are dedicated to the pursuit of pure knowledge and understanding.

To understand those differences and how they affect your research, imagine three conversations about balloons, dirigibles, zeppelins, and blimps.

For Entertainment. This conversation is conducted by people who meet to talk about zeppelins as a hobby. To join their conversation, you have only to show an interest in the topic and have a new fact or anecdote to offer—you bring a letter from Uncle Otto in which he describes his trip on the first zeppelin to cross the Atlantic and the menu from the diner. At stake here is a diverting hour with others who like to talk about zeppelins, and perhaps some personal enrichment. Your talk would be like the kind of paper you may have written in a composition class, one in which you were expected to be vivid, probably anecdotal, perhaps more interested in your own responses to a topic than in a dispassionate analysis of it. Since your task is to share with others your enthusiasm in a shared topic and to offer something those others didn't know and would find interesting, you will read your sources watching for amusing anecdotes, odd facts, and so on.

For Practical Benefit. Now imagine a second conversation, this one with people in the Public Relations department of Giganto, Inc. They would like to use a blimp to advertise Giganto, but they do not know how much it will cost or how effective it will be. They have employed you to find out. To succeed in this conversation, you have to understand that more is at stake than satisfying curiosity. You must answer a research question in a way that will help the PR people solve a practical problem by *doing* something: If they rent the blimp, does Giganto increase sales? That is the kind of audience you might write for when your teacher creates a "real world" scenario for your assignment—someone interested in *using* your research to solve a tangible, pragmatic problem in the world. If you know what these readers will do with your answers, you can know what information to seek out and what not to bother with—Giganto is unlikely to care about the first invention of a lighter-than-air craft or the equations for analyzing its aerodynamic stability.

For Understanding. Finally, imagine that your school has a Department of Lighter-than-Air Craft, with the same standing as the Department of English or Chemistry. The faculty teach about blimps, balloons, and zeppelins, research them, and participate in a worldwide conversation by publishing research on them. This conversation is carried on by hundreds, perhaps thousands of researchers. Some of them know one another, others have never met, but all of them read the same books and journals. Their objective is not to amuse themselves (though they do) or to help someone *do* something—like improve some company's image (though they would be pleased to serve as paid consultants to Giganto, Inc.). Their objective is to pose and answer questions about lighter-than-air craft, about their history, social consequences, literature, and theory. They determine the value of their work not by how well they entertain anyone or help anyone *do* anything, but by how much they learn and understand about blimps, by how close they can get to the truth.

As a consequence, these lighter-than-air scholars are intensely concerned with the intellectual *quality* of their conversation: they expect one another to be objective, rigorously logical, faithful to the evidence, able to see every question from all sides, regardless of where the inquiry leads or how long it takes. They expect the conversation to focus on complexities, ambiguities, uncertainties,

puzzles, and then to resolve them. They rely on one another's
research at the same time that they compete to produce their own:
so they test everything before they report it, because their central
value is to get things right and because they know that truth is
always partial—both incomplete and partisan. They understand
that whatever truth they offer is contestable and will be tested by
others in the conversation, not just to be contentious (though they
may be) or even nasty (though some will be), but to get closer to
the truth about blimps.

Such readers will take an interest in anything new you have to
say, but they will want to know what they should make of your
new information, how you think it affects what they *already* know
or understand about blimps. They will be especially interested if
you can convince them that they do not understand something as
well as they thought they did: *Most people think that lighter-than-air
craft originated in Europe in the eighteenth century, but I have discovered
a drawing of what appears to be a hot air balloon four centuries earlier, on
a wall in Central America.*

That is the kind of conversation you join when you report
research to a community of scholars. *Never mind whether your style is
graceful (though I will admire your work more if it is), never mind telling
me amusing anecdotes (though I like reading them if they help me understand
your ideas better), never mind whether what you know will make me rich
(though I would be happy if it did). Just tell me something I don't know so
that I can understand better what I do.*

These three kinds of readers may all be interested in lighter-
than-air craft, but their interests in them are different, and so they
will want your research to solve different kinds of *problems:* to
entertain them, to help them solve some problem in the world, or
simply to help them understand something better.

If this is your first foray into research, you must find out what
is at stake in your particular scenario. If you don't know, ask,
because these requirements will direct you toward different paths
of research.

Of course, in the process of your research, you might discover
something that will change your intention: as you are assembling
amusing anecdotes about the development of the zeppelin, you
discover that its standard history is wrong. But without some sense

of what you are up to from the beginning, you risk an aimless stroll through sources that will lead you—and your readers—nowhere in particular.

2.3 READERS AND THEIR COMMON PROBLEMS

Depending on what is at stake, readers and writers play different social roles, but behind those different roles are some concerns common to all readers and some problems common to all writers.

2.3.1 Readers and What You Know about Them

All readers share one interest: they want to read reports that impose on them as little unnecessary difficulty as possible. They may appreciate elegance and wit, but first they want to understand the point of your report and how you reached it. And so while it helps to think of the process of writing your paper as a path to a destination, it is also useful to imagine a similar journey for your readers, with you as their guide. They want your introduction to get them off on the right foot with a sense of where they are going and why you want to take them there, an idea of what question the journey will answer, what problem, scholarly or practical, it will solve.

Your readers will also want to know how you think your research and conclusions will change their thinking and beliefs: that's how they will gauge the *significance* of your report. Will you offer to a grateful audience the solution to a problem that they have long felt needed solving, or are you trying to sell a solution to readers who not only might reject it, but may not even want to hear about your problem in the first place?

All readers bring to a research report their own preconceptions and interests. So before you write, you have to think about where they stand and where you stand in regard to the question you are answering, and to the problem you are solving. If your question is already a live issue in your community, most readers will care about it before you pose it. In that case, you can focus on where they stand in respect to your answer:

- If they already know your answer, you are wasting their time.
- If they believe in a wrong answer, or in a right answer for the

wrong reasons, you first have to disabuse them of their error and then convince them that your answer is the right one for the right reasons—a difficult task.

• If they do not know an answer, you are in luck: you have only to convince them that you have the right answer and they will receive it with gratitude.

If, on the other hand, your question is not a live issue, your task is more complicated, because most readers will not be aware of your question or problem before you pose it. If so, you first have to convince them that your question is a good one.

• Some readers will have no interest in your question for any reason, so they will not care about your answer. Convincing them to care about your question in the first place may be a bigger challenge than convincing them you have answered it correctly.

• Some readers may be open to your problem because they see how its solution will help them understand their own problems better. If so, you are in luck.

• Other readers may reject your question and answer because to accept them would unsettle their understanding of long-held beliefs. They might change their minds, but only for good reasons powerfully stated.

• Finally, some readers are so entrenched in their beliefs that nothing will move them to think about a new question or an old problem in a new way. Such rare creatures you can only ignore.

2.3.2 Readers and What You Ask of Them

To understand your readers, then, you must know where they stand. But you must also decide where you want to take them and what they should do when they get there. It might be one or all of the following.

Accept New Knowledge. If you offer readers only new knowledge and new conclusions, you must either assume that they have a prior interest in your topic or convince them it is in their interest to make mental space for it. If they have a prior interest, just offering information is least demanding; it is also least interesting and found least often. Occasionally a researcher will say, in effect, *Here is some information I have uncovered that I hope someone might find* *interesting.* Readers already interested might be grateful, but they will be more interested if the researcher can show how the new data force them to entertain a new question, especially if the new data disturb old ways of thinking.

If you find yourself with material on nineteenth-century Tibetan weaving that might be new to your readers but you have no point other than *You probably don't know about this,* that's OK. Better would be to think hard about how your new information might require them to think differently about Tibet, about weaving, or even about the nineteenth century. That means finding questions that might interest your readers and that your new knowledge can answer.

In the world of business and commerce, it is common for a supervisor to direct researchers just to gather and report information, but that person usually wants the information to solve a problem she already knows she has. In that case, there is a division of labor: *You get me the information I need so that I can solve my problem.*

Change beliefs. You ask more of your readers (and yourself) if you ask them not just to make room for new knowledge but to change deeper beliefs as well. The more deeply they hold their beliefs, the more it takes to change them. That's how readers measure significance. For example, it would take little to convince most of us that there are exactly 202 asteroids known to be a mile or more across, because few of us care. But if we could be convinced that those 202 asteroids were remnants of a planet that once existed between Earth and Mars but was blown up in a nuclear war, we would have to change many beliefs about many important things, the least of which would be the exact number of asteroids. When you think about the point you are making, think as well about the impact that you want it to have on the general structure of your reader's beliefs and understanding. The greater the impact, the more significant your point and the harder you have to work to be convincing.

The painful fact, though, is that even experienced researchers find it difficult to anticipate how their findings will cause readers to change their beliefs. And even when they do understand, they often struggle to explain why their readers should change.

Now this is important: *If you are a beginning researcher, do not think that you have to meet an expectation that high.*

You don't have to worry, at first, whether the results of your research will be new to others or will change anyone's mind *but your own*. Be concerned first with whether your work is important to *you*. If you can find a question that *you alone* want to answer, you have achieved something substantial. If you can find an answer that changes only what *you* think about a good many things, you have achieved something even more significant—you have discovered how new ideas unsettle and rearrange stable beliefs.

If you are an advanced researcher, however, you *must* take the next step. Your readers expect you to pose a problem that they recognize not just as *your* problem, but as *their* problem as well, a problem whose solution will change *their* thinking in ways *they* think significant. (We'll address this demand in more detail in Chapter 4.)

Perform an Action. Occasionally, researchers ask readers to perform an action because they believe that the solution to their research problem can help their readers solve a tangible, practical problem in the world. Sometimes this is easy—a chemist figures out how to produce pollution-free gasoline and then tries to persuade oil companies to use his formula.

More often, the results of your research will point not to a specific action but rather to a conclusion that will change only your readers' understanding. In the world of scholarly research, though, that is no small achievement. In the final analysis, the significance of academic research hinges on how much it changes and rearranges belief, regardless of whether those new beliefs lead to an action.

Keep in mind that just about every academic researcher began by satisfying interests, not of his readers, but of his own. Also be aware that even experienced researchers often cannot answer these questions about the significance of their research right from the start. As paradoxical as it may sound, not until they have finished a first draft of their report do most of them fully understand the significance their findings should have for others. So here is another piece of reassuring advice to those beginning their first project: When you begin with your own interests—*as you should*—you will probably not know what you will ask of your readers, or even of yourself. You will discover that only after you have found an answer that helps you understand better the question you want to pose and your readers to consider. Even then, your best reader may be yourself.

> Nothing is more important to successful research than your own commitment to it. Some of the world's most important research has been done by those who prevailed in the face of indifference, because they never doubted their own vision. Barbara McClintock, a geneticist, struggled for years, unappreciated because her research community did not consider her work significant. But she believed in it, and eventually, when her community was persuaded to ask questions that only she could answer, she won science's highest honor, the Nobel Prize.

2.4 WRITERS AND THEIR COMMON PROBLEMS

Just as all readers share certain concerns, so all writers face some of the same problems. The most important one for beginners is the difference that experience makes. When writers know a field, they internalize its practices so well that what they once could do only by rule and reflection they now do by habit. Practiced writers begin with an intuitive sense of what readers expect, of what will be the final shape of their paper. Less experienced writers have to concentrate not only on their specific issues and problems but also on what experienced writers do intuitively. But of course that is why you are expending all this effort in the first place, to learn how to do more research with less wasted effort. And that is the goal of this book: to offer you guidelines, checklists, and quick tips that help you evaluate your progress and your plans and, most important, to show you how to *think and write like a reader:* in short, to make explicit what experienced writers do intuitively.

Everyone starts as a novice, and most of us feel like novices again when we begin a new project about which we are not entirely confident. We three authors remember trying to write up preliminary conclusions, aware that our writing was uncertain and confused because we were too. We remember just repeating what we had read when we should have been analyzing, synthesizing, and criticizing it. We had that experience when we were students, first as undergraduates and then in graduate school, and we have it again just about every time we begin a project that requires us to explore something genuinely new.

As your skills and experience grow, you will overcome some of those anxieties. Practice pays off. Why, then, once you have "learned how to do research" can you not escape them entirely? The fact is that learning to do research is not like learning to ride a bike, a skill that you can repeat each time you try a new one. Doing research does involve some repeatable skills, but since the subjects of research are infinitely various, and the ways of reporting results vary from field to field, each new project brings with it new problems. The difference between expert and novice is partly that the expert controls the repeatable skills better, but also that she can better anticipate the inescapable uncertainties and deal with them.

So how do you avoid feeling overwhelmed?

First, be aware of those uncertainties that you will inevitably experience. That should be the object of your first quick reading of this book.

Second, master your topic by writing about it *along the way.*

> ### Cognitive Overload: Some Reassuring Words
>
> The difficulties that beginning researchers experience have less to do with age or achievement than with their experience in a particular field. One of us was once explaining to teachers of legal writing how the problems of being a novice create feelings of insecurity in new law students, even among those who were good writers as undergraduates. At the end of the talk, one woman announced that when she began law school, she experienced some of those same feelings of uncertainty and confusion. Before law school, she had been a professor of anthropology, had published her work and been praised by reviewers for the clarity and force of her writing. Then she switched careers and went to law school. She said that during her first six months she wrote so incoherently that she began to fear that she was suffering from a degenerative brain disease. She wasn't, of course: she was only experiencing a kind of temporary aphasia that afflicts most of us when we try to write about matters we do not entirely understand. Not surprisingly, the better she began to understand the law, the better she began to think and write.

Don't just photocopy sources and underline words: write summaries, critiques, questions to think about later. The more you write *as you go,* no matter how sketchily, the more confidently you can face that intimidating first draft.

Third, control the complexity of your task. All parts of the research process affect all the other parts, so use what you learn about the parts to unbundle this complex collection of tasks into manageable steps. Manage the first steps in finding a topic and formulating some good questions, and you'll work more effectively later when you begin to draft and revise. Conversely, if you can anticipate how you will later draft and revise, you can more effectively look for a topic and formulate a problem now. You can give each task the attention it requires if you understand the tasks you face—how to coordinate them, when to concentrate on particular ones, when to stand back and take stock, how to review your plans, even when to change them.

Fourth, count on your teacher to understand your struggles. Good teachers want you to succeed, and you can expect their help.

Most importantly, recognize the problem for what it is: your struggles need not signal any deep failing on your part. To get over the problems that all beginners face, do exactly what you are doing, what every successful researcher has done: Press on.

Quick Tip:
A Checklist for Understanding Your Readers

Although you should think about your readers right from the start, you can't expect to answer all of the following questions until you near the end of your research. So plan to return to this checklist a few times, each time filling out more of the picture of the role you will create for your readers.

Who Is Your Community of Readers?

1. Are your readers
   - Professionals in the field of your research?
   - General readers who have
     −different levels of knowledge and interest?
     −similar levels of knowledge and interest?
2. For each uniform group of readers, repeat the following analysis.

What Do They Expect You to Do for Them?

1. Entertain them?
2. Help them solve some problem out there in the world?
3. Help them understand something better?

How Much Do They Know?

1. Level of background knowledge (compared to you):
   much less   less   same   more   much more

2. Knowledge about the particular topic (compared to you):
   much less   less   same   more   much more

3. What special interest do they have in this topic?
4. What matters do they expect you to discuss about this topic?

Do They Already Understand Your Problem/Question?

1. Is the problem of this paper one that your readers recognize?
2. Is it one that they have but haven't yet recognized?
3. Is the problem not theirs, but yours?

4. Will they immediately take the problem seriously, or must you persuade them that it matters?
5. Is the research problem motivated by a tangible difficulty in the world or by a scholarly, conceptual difficulty?

How Will They Respond to Your Solution/Answer?

1. What do you expect readers to *do* as a result of reading your report: accept new information, change certain beliefs, take some action?
2. Will the solution contradict what they already believe? How?
3. Will readers already know some standard arguments against your solution?
4. Will the solution stand alone or will readers want to see the steps that led to it?

In What Forum Will They Encounter Your Report?

1. Have your readers asked for your report? will you send it to them unbidden? will they encounter it in a publication?
2. Before it reaches your main readers, will your report have to be approved by a gatekeeper—your supervisor, an editor of a publication, an assistant to an executive or administrator, a technical expert?
3. Do readers expect your report to follow a standard format? If so, what is it?

PART TWO

Asking Questions, Finding Answers

Prologue: Planning Your Project

IF YOU HAVE SKIMMED THIS BOOK ONCE, you are ready to begin your project. But before you head to the library, you have to do some careful planning. If you have an assignment that defines a question and specifies each step in your project, skim the next two chapters again, follow the instructions in your assignment, then return to Part III before you start drafting. If on the other hand you have to plan your own research, even find your own topic, you may feel daunted. But your task is manageable if you tackle it a step at a time.

No single formula can guide everyone's research: you'll spend time searching and reading just to discover where you are and where you are going; you'll spend time in blind alleys; and you'll learn more than your paper requires. In the end, however, that extra work will pay off, not just in a good paper, but in your ability to deal with new problems more effectively.

As you begin, anticipate that you will have to take these first steps:

- You must settle on a *topic* specific enough to let you master a reasonable amount of information: not "the history of scientific writing," but "essays in the *Proceedings of the Royal Society* (1800–1900) as a precursor to the modern scientific article."
- Out of that topic, you must develop *questions* that will guide your research and point you toward a *problem* that you intend to solve.
- You must gather *data* relevant to answering your question.

When you have collected data that answer most of your questions, you then of course have to shape them into an argument (the topic of Part III) and draft it (the topic of Part IV).

As you collect, sort, and assemble your information, plan to do lots of writing. Much of it will be simple note-taking just to record what you've found, but it should also include "writing to understand": outlines, diagrams of how seemingly disparate facts relate, summaries of sources and "positions" and "schools," lists of related points, disagreements with what you have read, and so on. While little of this preliminary writing will end up in your final draft, it is important to do it, because writing about your sources *as you go* helps you better understand them and encourages your own best critical thinking. It will also help you when you sit down to begin your first draft.

> ### What Are Your Data?
>
> No matter what their field, all researchers generate information as evidence to support their claims. But researchers in different fields call evidence by different names. Since the most common is *data*, we will use that term to refer to all the kinds of information used in different fields. Be aware that by *data* we mean more than the quantitative information common in the natural and social sciences, even though the term may jangle in the ears of researchers in the humanities.

You will discover quickly that you cannot move through these steps in the neat order that we present them. You will find yourself drafting a summary before you have gathered all your data; you will start formulating an argument before you have all your evidence; and when you think you have an argument worth making, you may discover you have to return to the library for more evidence. You may even discover that you have to rethink the questions you have been asking. Doing research is not a process in which you can move from here to there in a neat, linear way. But however indirect your progress may be, you will feel more confident that you are in fact progressing if you can understand and manage the components of the process.

Quick Tip:
Writing in Groups

We will suggest that you seek out friends to read versions of your report so that you can better see it as others will. But you may also be asked to write a report as part of a group effort. If so, you face both opportunities and challenges: A group has more resources than someone working alone, but to profit from that advantage the group must manage itself thoughtfully.

Three Keys for Working in Groups

Talk a Lot

The first key to writing in groups is to talk a lot and agree on a plan. Even more than does a single writer, a group needs a plan, and talking is the only way to create one, to monitor your progress, and more importantly to change that plan as you understand your project better. Set regular meeting times, hold weekly conference calls, share e-mail addresses—do whatever you can to ensure that everyone talks with one another at every opportunity.

Before you start, be sure that your group agrees on goals—on the question or problem you address, the kind of claim you expect to offer, the kind of evidence you will need to support it. Your group will modify these goals as you understand your project better, but you must start with some shared understanding. Your group should discuss readers—what they know, what's important to them, what you expect them to do with your report. Finally, the group should lay out the steps that will achieve your goal—exactly who will do what and when.

To focus talk on the stages of your project, use these chapters and sections as a guide. Use checklists to share ideas about readers (pp. 26–27), to ask questions systematically (pp. 39–41), to reformulate them as a problem (pp. 52–56). Assign someone to maintain a common outline that is updated regularly, first as a topic outline (p. 152), then as an outline of your argument (p. 106), and finally of your points (p. 153). If your project involves lots of data, create a schedule to gather them; maintain a list of sources consulted and still to be consulted, with brief annotations on how useful each source has been.

The more you and your group talk together, the more easily

you will write together. If, as with the three authors of this book, the members of your group share background knowledge, have worked together before, and can anticipate what one another thinks, you can talk less. Even so, in writing this book, we three made scores of phone calls, exchanged hundreds of e-mail messages, and sat together a dozen times (sometimes driving more than a hundred miles to do so).

Agree to Disagree, and Then to Agree

Sharing is essential, but don't expect your group to agree 100% on every issue. You can expect to differ over particulars, sometimes quite a bit. As you resolve those differences, your group can produce its best thinking, because you have to be explicit about what you believe and why. On the other hand, nothing impedes progress more than someone's insisting on *his* wording, on including *her* bit of data. If the first rule of writing in a group is to talk a lot, the second is to keep disagreements in perspective. When you disagree over issues that have no significant impact on the whole, forget it. Reserve your intransigence for matters of ethical principle or fundamental understanding.

Organize Yourselves into a Team with a Leader

The group should ask someone to serve as their moderator, facilitator, coordinator, organizer; there are different names for the job, but most groups need someone to keep track of the schedule, ask about progress, moderate discussions, and when the group seems deadlocked, decide which way to go. That job can rotate among the group, or one person can hold it for the duration of the project. The rest of the group simply agrees that after extensive discussion and debate, the moderator/facilitator makes a decision and everyone agrees and moves forward.

THREE STRATEGIES FOR WORKING IN GROUPS

Here are three ways groups can organize their work, and some of the risks of each. Most groups combine these strategies to fit their particular situation.

Divide, Delegate, and Conquer

This strategy exploits the fact that a group has more skills than an individual. It succeeds best when members have different backgrounds and talents, and the group divides the tasks to make the best use of each. A group working on a sociological survey, for example, might decide that two people are good at gathering data, two others at analyzing them and producing graphics, two more at drafting, and all will take a turn at editing and revising. This strategy depends on each participant reserving enough time for his or her job in the sequence when it has to be done. If others have less to do at any given point, they can do legwork as needed.

The *least* successful use of this strategy is to divide the document into sections for each member to research, organize, draft, and revise. This works only when the parts of a report are relatively independent. But even then someone will have to make them hang together, and that can be nasty work, particularly if the members of the group have not consulted along the way.

No matter how the group divides the work, it needs good management skills, because the greatest danger is lack of coordination. Whether you parcel out tasks or parts, you should spend time talking about what you are doing and being utterly clear about who is supposed to do what. Then write it down and give everyone a copy.

Write Side-by-Side

Some groups share all the work, working side-by-side the whole way. This strategy works best when a group is small, tightly-knit, can work well together and spend a lot of time on the task—for example, a group of engineering students who devote two semesters to one design project. The disadvantage is that some people are uncomfortable talking about half-formed ideas before they work them out in writing. Others find it even more uncomfortable to share rough and unrevised drafts. The members of a group that uses this strategy must be tolerant with one another. What often happens is that the most confident person in the group will ignore the feelings of the others, dominate the process, and inhibit progress.

Take Turns

Some groups share all of the work, but draft and revise in turns, so that a text evolves toward a final version as a whole.

This strategy is effective when members of the group differ about what is important, but those differences complement rather than contradict one another.

For example, in a group working on stories of the Alamo, one person might be interested in the clash of cultures, another in political consequences, and a third in the role of narrative in popular culture. The members might work from the same sources but identify different issues as important to their own perspectives. But then, after sharing what they have found, the members of the group take turns working on versions of a whole draft. The first writer creates a rough and incomplete draft, but with enough structure for others to see the shape of the argument and to expand and rearrange. Each member in turn then takes over the draft, adding and developing ideas that seem most important. The group agrees that the person working on the draft at that moment "owns" it and can make whatever changes that person wishes, so long as the changes reflect the group's understanding of the whole.

The risk is that the final product will seem to work at cross-purposes, following a zigzag path from one incompatible interest to another. A group that works by turns has to agree on the final goal and the shape of the whole, and each member has to respect and accept the perspectives of the others.

Your group may find that it can use different strategies at different stages. For example, in early planning, you may want to work side-by-side, at least until you form a general sense of your problem. For data-gathering, you may find it most efficient to divide and conquer. And for the final stages of revision, you may want to take turns. In writing this book, we mixed strategies. Early on, we worked side-by-side until we had an outline. We then worked on separate chapters and returned to work side-by-side when our progress stalled and we felt that we had to revise our plan (which happened at least three times). Most often, though, we divided the work by each drafting individual chapters. Once we had a whole draft, we worked by turns, with the result that many of these chapters resemble very little the chapter that one or the other of us originally drafted.

Working in groups is hard work, and sometimes hard on the ego, but it can also be highly rewarding.

CHAPTER THREE

From Topics to Questions

*In this chapter, we discuss how to use your interests to find a topic,
narrow it to a manageable scope, and then generate questions that will
focus your research. If you are an advanced student and already have a
dozen topics that you would like to pursue, you might skip to Chapter
4. If, however, you are starting your first project, you will find this
chapter useful.*

3.1 INTERESTS, TOPICS, QUESTIONS, AND PROBLEMS

IF YOU ARE FREE to pursue any research topic that interests
you, that freedom may be frustrating—so many choices, so little
time. Finding a topic, though, is only the first step, so do not
assume that once you have a topic, you need only search for information and report what you find. Beyond a topic, you have to find
a reason (other than completing your assignment) for devoting
weeks or months to pursuing it and then for asking readers to
spend time reading about it.

Researchers do more than just dig up information and report
it. They *use that information to answer a question that their topic inspired
them to ask.* At first, the question may intrigue the researcher alone:
how good was Abe Lincoln at math? Why do cats rub their faces
against us? Is there such a thing as innate perfect pitch? That's how
most significant research begins—with an intellectual itch that only
one person feels the need to scratch. But at some point, a researcher
has to decide whether the question and its answer might be *significant,* at first to the researcher alone, but eventually to others—to
a teacher, to colleagues, to an entire community of researchers.

At that point, the researcher must view his task differently: he
must aim not just at answering a question, but at posing and solving
a *problem* that he thinks *others* should also recognize as worth solving.
That word "problem," though, has a meaning so special in the
world of research that it is the topic of the whole next chapter. It
raises issues that few beginning researchers are ready to resolve
entirely, and that can vex even an advanced researcher. So do not

feel dismayed if at first you cannot find in your topic a problem
that others might think worth solving. But you will never even
approach that point unless you strive to find in your topic a question
that at least *you* think worth asking.

In this chapter, we focus on the steps leading to the formulation
of a research question. How do you transform an interest into a
topic for research? How do you find questions that can guide your
research? Then how do you decide whether those questions and
answers are worth pursuing, at first just to you, but then to your
readers? The process looks like this:

1. Find an interest in a broad subject area.
2. Narrow the interest to a plausible topic.
3. Question that topic from several points of view.
4. Define a rationale for your project.

In the next chapter we address the more vexing matter of turning
your questions into a research *problem*.

3.2 FROM AN INTEREST TO A TOPIC

Experienced researchers have more than enough *interests* to pursue. An interest is just a general area of inquiry that we like to
explore. The three of us have our current favorites: society and
language, textual coherence and cognition, ethics and research. But
while beginning researchers also have interests, they sometimes find
it difficult to locate among them a *topic* appropriate for academic
research. A topic is an interest specific enough to support research
that one might plausibly report on in a book or article that helps
others to advance their thinking and understanding: the linguistic
signals of social change in Elizabethan England, the role of mental
scenarios in the reader's creation of coherence, the degree to which
current research is motivated by under-the-counter payments.

If you are free to explore any topic within reason, we can offer
only a cliché: start with what interests you most deeply. Nothing
will contribute to the quality of your work more than your sense
of its worth and your commitment to it. Start by listing four or
five areas that you'd like to learn more about, then pick one with
the best potential for yielding a topic that is specific and that might
lead to good sources of data. If you are in an advanced course, you
are likely to be limited to matters of interest to those in your field
of study, but you can always find more by looking in a recent

textbook, talking to another student, or consulting your teacher. You might even try to identify an interest that will provide a topic for work in another course, either now or in the future.

If you are still stuck, here is a way to search for topics that might pan out: If this is your first research project in a writing course, find in the reading room of your library a general bibliographical resource such as the *Reader's Guide to Periodical Literature* or the *Bibliographic Index* (we will discuss these resources in more detail in Chapter 5 and in the Quick Tip after it). If you are an advanced student, locate a specialized index in your particular field, such as *Philosopher's Index, Psychological Index, Women's Studies Abstracts.* Now skim its headings until you find one that catches your interest. That heading will provide not only a possible topic, but also a list of sources on it.

If you are writing your first research paper in a particular field and have not yet settled on a topic, you might head over to the library to find out where its resources are particularly rich. If you pick your topic first and then after considerable searching discover that the sources are thin, you will have to start over. By identifying areas with promising resources, you learn the strengths and weaknesses of your library and can plan this and future projects more thoughtfully. (If you are really stuck, look at the Quick Tip at the end of this chapter for more suggestions.)

3.3 FROM A BROAD TOPIC TO A NARROW ONE

At this point, you risk picking a topic so broad that it could be a subheading in an encyclopedia article: "Space flight, history of"; "Shakespeare, Problem Plays"; "Natural kinds, doctrine of." A topic is probably too broad if you can state it in fewer than four or five words. If you find yourself struggling with that kind of topic, narrow it:

Free will and historical inevitability in Tolstoy's *War and Peace* →	The conflict of free will and historical inevitability in Tolstoy's description of three battles in *War and Peace*
The history of commerical aviation →	The contribution of the military to the development of the DC-3 in the early years of commercial aviation

We narrowed these topics by adding modifying words and phrases. In particular, we added four nouns of a special kind: *conflict, description, contribution,* and *development*. Those nouns are special because they are each related to a verb: *conflict, describe, contribute,* and *develop*. At some point, you will have to move from a phrase that names a topic—"free will and historical inevitability in Tolstoy," "history of commercial aviation"—to a sentence that states a potential *claim*. If you narrow your topic by using nouns derived from verbs, you will be one step closer to a claim that could be challenging enough to interest your readers. Compare these:

Free will and historical inevitability in Tolstoy's *War and Peace*	→	There is both free will and historical inevitability in Tolstoy's *War and Peace*.
The **conflict** of free will and historical inevitability in Tolstoy's **description** of three battles in *War and Peace*.	→	Tolstoy **describes** three battles in a way that makes free will **conflict** with historical inevitability.
The history of commercial aviation.	→	Commercial aviation has a history.
The **contribution** of the military in the **development** of the DC-3 in the early years of commercial aviation.	→	The military **contributed** to the way the DC-3 **developed** in the early years of commercial aviation.

These may not be particularly interesting claims yet. But since you will build your final project out of a series of claims, you should, from the beginning, take every opportunity to work toward the kinds of claims you will eventually need.

The advantage of a specific topic is that you more easily recognize gaps, inconsistencies, and puzzles that you can question. That will help you turn your *topic* into a research *question*. (If you follow our later suggestion to begin with an index or abstract, your topic will already be restricted by its headings.)

Caution: you narrow your topic too severely when you cannot easily find sources.

The history of commercial aviation
↓
Military support for development of the DC-3 in the early years of U.S. commercial aviation
↓

The decision to lengthen the wing tips on the DC-3 prototype as a result of the military desire to use the DC-3 as a cargo carrier

3.4 FROM A NARROWED TOPIC TO QUESTIONS

Once the beginning researcher hits on a topic that feels *both* interesting and promising, perhaps something like "the political origins and development of legends about the Battle of the Alamo," she typically begins searching out sources and collecting information, in this case versions of the story in books and films, Mexican and American, nineteenth century and twentieth. She might then write a paper that summarizes the stories, points out differences and similarities, contrasts them with what modern historians think really happened, and concludes,

Thus there are interesting differences and similarities between . . .

In a first-year writing course, such a paper might earn a passing grade. It shows that the student can focus on a topic, find data on it, assemble those data, and present them coherently—no small achievement for a first research project. But for anyone who wants her research to *matter,* such an achievement would fall short of the mark.

While the writer may have learned something from the exercise of searching out and reporting on the Alamo stories, she offers only *information.* She asked no *question* that she or her readers might think worth asking, and so she can offer no *answer* significant enough to change how she or her readers should think about those stories or their development.

Once you have a topic to research, you should find in it questions to answer. Questions are crucial, because the starting point of good research is always what *you do not know or understand but feel you must.* Start by barraging your topic with question after question, first with the obvious standing questions of your field:

> Do the legends about the Battle of the Alamo accurately reflect our best historical accounts? Do the historical accounts differ?

Ask the standard *who, what, when,* and *where* questions. Record your questions, but don't stop for their answers.

You can organize your questions from these four perspectives:

1. What are the parts of your topic and what larger whole is it a part of?
2. What is its history and what larger history is it a part of?
3. What kinds of categories can you find in it, and to what larger categories of things does it belong?
4. What good is it? What can you use it for?

(Don't worry about getting the right questions in the right categories; the categories serve only to stimulate the questions.)

3.4.1 Identify Its Parts and Wholes

• Question your topic in a way that analyzes it into its component parts and evaluates the working relationships among them:

> *What are the parts of stories about the Battle of the Alamo? How do they relate to one another? Who were the participants in the stories? How do the participants relate to the place, the place to the battle, the battle to the participants, the participants to one another?*

• Question your topic in a way that identifies it as a working component in a larger system:

> *What use have politicians made of the story? What role does it have in Mexican history? What role does it have in our history? Who told the stories? Who listened? How does the nationality of the teller affect the story?*

3.4.2 Trace Its History and Changes

• Question your topic in a way that treats it as a dynamic entity that changes through time, as something with its own history:

> *How did the battle develop? How have the stories developed? How have different stories developed differently? How have audiences changed? How have the storytellers changed? How have motives to tell the story changed? Who first told the stories? Who told them later? Who were the earliest readers and listeners? Who later?*

• Question your topic in a way that identifies it as an episode in a larger history:

What caused the battle, the stories? What did the battle and the stories then cause? How do the stories fit into a historical sequence? What else was happening when the stories appeared? When they changed? What forces caused the story to change?

3.4.3 Identify Its Categories and Characteristics

- Question your topic in a way that defines its range of variation, how instances of it are like and different from one another:

  What is the most typical story? How do other stories differ from it? Which one is most different? How do the written and oral stories differ from the movie versions? How are Mexican stories different from ours?

- Question your topic in a way that locates it in a larger category of things like it:

  What other stories in our history are like the story of the Battle of the Alamo? What other stories are very different? What other societies have the same kinds of stories?

3.4.4 Determine Its Value

- Question your topic in regard to the value of its uses:

  What good are the stories? What use has been made of them? Have they helped people? harmed them?

- Question your topic in regard to the relative value of its parts and features:

  Are some stories better than others? What version is the best one? the worst one? Which parts are most accurate? Which least?

3.4.5 Review and Rearrange Your Answers

When you run out of questions, group them in different ways. In the Alamo example some questions relate to the development of the stories; others address their quality as fact or fiction; others highlight differences between versions (nineteenth- and twentieth-century, Mexican and American, written and movie); other questions address political issues, and so on. Such lists can provide scores of research topics. If they are freewheeling enough, they can have the exhilarating effect of opening up worlds of research.

The next step requires more careful judgment. First, identify questions that need more than a one- or two-word answer. Questions that begin with *who, what, when,* or *where* are important, but they ask only about matters of fact. Emphasize instead questions that begin with *how* and *why*. Then decide which questions stop you for a moment, challenge you, spark some special interest. At this point, of course, you can't be sure of anything. Your answers may turn out to be less surprising than you hoped, but your task now is only to formulate a few questions whose answers *might* be both plausible and interesting.

When you've done all this, you have taken your first big step toward a project that goes beyond just collecting data. You have identified something that you don't know but want to, and what you want to know drives the earliest stages of your research. You are ready to gather data, a process we'll describe in Chapter 5. But even though you can now begin gathering data, the process of focusing your project is not yet complete.

3.5 FROM A QUESTION TO ITS SIGNIFICANCE

Even if you are an experienced researcher, you may not be able to take this next step until you are well into your project, perhaps even close to its end. And if you are a beginning researcher, you may feel this step is especially frustrating. Once you have a question, you have to ask and try to answer the further question, *So what?*

> So what if I don't know or understand how snow geese know where to go in the winter, why the Titanic was designed so badly, how fifteenth-century violin players tuned their instruments, why Texans tell one story about the Alamo, Mexicans another? So what?

This question vexes all researchers, beginners and experienced alike, because to answer it, you have to know how significant your research might be not just to yourself but to others. Instead of asking that question straight out, though, you can get closer to its answer if you move toward it in steps.

3.5.1 Step 1: Name Your Topic

In the earliest stages of a research project, when you have only a topic and maybe the first glimmerings of a few good questions, try to describe your work in a sentence something like this:

I am learning about/working on/studying _____.
Fill in the blank with a few noun phrases. Be sure to include one or two of those nouns that you can translate into a verb or adjective:

I am studying *repair processes* for cooling systems.

I am working on the *motivation* of President Roosevelt's early speeches.

3.5.2 Step 2: Suggest a Question

As early as you can, try to describe your work more exactly by adding to that sentence an indirect question that specifies something about your topic that you do not know or fully understand, but want to:

I am studying X *because I want to find out* who/ what/ when/ where/ whether/ why/ how _____.

You now have to fill in the new blank with a subject and a verb:

I am studying repair processes for cooling systems, *because I am trying to find out how* expert repairers analyze failures.

I am working on the motivation of Roosevelt's early speeches, *because I want to find out whether* presidents since the '30s have used those speeches to announce new policy.

When you can add that kind of *because-I-want-to-find-out-how/ why* clause, you have defined both your topic and your reason for pursuing it. If you are doing one of your first papers and you get this far, congratulate yourself, because you have defined your project in a way that goes beyond the random collection of information.

3.5.3 Step 3: Motivate the Question

There is, though, one more step. It's a hard one, but if you can take it, you transform your project from one that interests you to one that makes a bid to interest others, a project with a rationale explaining why it is important to ask your question at all. To do that, you must add an element that explains why you are asking your question and what you intend to get out of its answer.

In Step 3, you add a second indirect question, this one introduced with *in order to understand how, why,* or *whether:*

1. I am studying repair processes for cooling systems,
   2. because I want to find out how expert repairers ana-
      lyze failures,
      3. *in order to understand how* to design a computerized
         system that could diagnose and prevent failures.

1. I am working on the motivation of Roosevelt's early
   speeches,
   2. because I want to discover whether presidents since
      the '30s used those speeches to announce new
      policy,
      3. *in order to understand how* generating public support
         for national policy has changed in the age of tele-
         vision.

Assembled, the three steps look like this:

1. *Name your topic:*

   I am studying _____,

   2. *Imply your question:*

   because I want to find out who/how/why _____,

      3. *State the rationale for the question and the project:*

      in order to understand how/why what _____.

Rarely can a researcher flesh out this pattern fully before she begins
gathering information. In fact, most can't complete it until they're
nearly finished. Too many, unfortunately, publish their results without having thought through these steps at all.

Even though at the beginning of your project you won't be
able to state these steps fully, it is a good idea to test your progress
every so often by seeing how close you can come. Better: Have
someone else—roommate, relative, or friend—*force* you to flesh
out this progression. Your evolving description will help you keep
track of where you are and keep you focused on where you may
still have to go.

It may be that in your first try at research you will not find a
question whose answer has much significance to anyone but yourself. But do that much and you will delight your teacher. As you
move through your project, though, do what you can to fill out
the pattern; try to find a reason for asking your question, a way to
make its answer seem *significant* to you, maybe even to others.

Remember, your eventual object is to explain,
- what you are writing about—your topic.
  - what you don't know about it—your question.
    - why you want to know about it—your rationale.

When you can achieve these three objectives, you will have articulated a motive for your project that goes beyond just meeting a requirement. You will know that you have an *advanced* research project when what follows the *in order to understand* is important not just to you but to your readers as well.

It is when we begin to consider our readers that we must change the terms of our project from posing and answering a question to posing and solving a problem, the subject of our next chapter.

Quick Tip:
Finding Topics

If you are an advanced researcher, chances are that you will not have to look far for topics to research. You can focus on current research in your field, which you can find easily enough by browsing through recent articles and review essays and, if they are available, recent dissertations, especially the suggestions for future research in their conclusions. If you are less advanced, your teacher will still expect you to focus your topics on the field, though not on its most advanced state. Most teachers will either assign topics to choose from or at least indicate the kind of topics to consider.

But sometimes you will be left to find topics on your own, and if you are in a first-year writing class, you will have to find good topics without even a specific field to focus your efforts. If you have to find your own topic and have drawn a blank, try looking in these sources:

For Topics Focused on a Particular Field of Study

1. Browse through a textbook in a course one level advanced beyond yours or from a course that you know you will have to take at some time in the future. Don't overlook the study questions.

2. Attend a public lecture in your field and listen for something you disagree with, don't understand, want to know more about.

3. Browse through the topic headings in specialized bibliographies and abstracts.

4. Browse through the *Encyclopedia of . . .* in the field you are studying.

5. Ask your instructor about the most contested issue in her field.

6. If you have access to the internet, find a specialized "list" that interests you and "lurk" (read messages sent by others) until you find debated topics.

For General Topics

1. Think of some special interest you have—sailing, gymnastics, chess, volunteer work, modern dance—and investigate its origins or how it is practiced in another culture.

2. Investigate a specific aspect of a country you'd like to visit.

3. Wander through a museum of any kind—art, natural history, automobile—until you find yourself looking at something with great interest. What more do you want to know about it?

4. Wander through a large shopping mall or store, asking yourself, *How do they make that?* or *I wonder who thought up that product?*

5. Leaf through your Sunday newspaper, especially the features sections, until you find yourself stopping to read something. If you have access to the *New York Times,* look through its feature sections and the Sunday review of books.

6. Go to a large magazine rack and browse. Buy a magazine that looks technical and interesting. Look especially for trade magazines or those that cater to highly specialized interests.

7. Look through the kind of popular magazines you find in waiting rooms, such as the *Reader's Digest,* for an article that makes a significant claim about health, society, or personal relationships that is based on alleged "evidence." Find out whether it is true.

8. Tune into interview programs on TV or talk radio until you hear something you disagree with. Then ask yourself whether you could find enough information to refute it.

9. Recall the last time you heatedly discussed some important topic and were frustrated because you didn't have the facts you needed.

10. Think of one thing that you believe but most people don't. Then ask whether it's the kind of issue on which you can find enough evidence to convince someone else.

11. Think of some common beliefs that everyone takes for granted but might not be so, such as the claim the Eskimos have scores of words for snow or that one gender is naturally better at something than the other.

12. Skim topic headings in general bibliographies, such as the *Readers' Guide to Periodical Literature.*

13. Think of a popular controversy that research could help you clarify.

14. Get together with five or six friends and brainstorm about what you would all like to know more about.

CHAPTER FOUR

From Questions to Problems

*This chapter covers matters that beginning researchers may find diffi-
cult, perhaps even baffling. So those of you working on your first project might skip to Chapter 5. (Of course, we hope that you will rise to
the challenge and read on.) For advanced students, though, what follows is essential.*

IN THE LAST CHAPTER, we described how to find in your
interests a topic, how to find in that topic questions to research,
and then how to signal the significance of your answer by describing
its rationale:

1. *Topic:* I am studying _____,
   2. *Question:* because I want to find out who/how/
     why _____.
    3. *Rationale:* in order to understand how/why/
      what _____.

These steps define not only the development of your project,
but your own growth as a researcher. When you move from step
1 to 2, you go beyond those who merely gather information, because
you are directing your project not by aimless curiosity (by no means
a useless impulse), but by your need to understand something better. When you move on to step 3, you surpass beginning researchers, because you are focusing your project on the *significance,* on
the *usefulness* of understanding what you do not know. When those
steps become a habit of thinking, you become a true researcher.

4.1 PROBLEMS, PROBLEMS, PROBLEMS

There is, though, a last step, one that is hard for even experienced researchers. You must convince your readers that the answer
to your question is significant not just to you, but to *them* as well.
You must transform your motive from discovering to *showing;* more
importantly, from understanding to *explaining* and *convincing.*

This last step trips up even experienced researchers, because they often think that they have done their job simply by posing and answering a question that interests them. They are only partly right: their answer must also be the solution to a *research problem* that is significant to others, either because those others already think it is significant or, as is more likely, because they can be convinced to think so. What sets you apart as a researcher of the highest order is the ability to develop a question into a problem whose solution is significant to your research community. The trick is to communicate that significance. To understand how to do that, you have to understand more exactly what we mean by a research "problem."

4.1.1 Practical Problems and Research Problems

Most everyday research begins not with finding a topic but with confronting a problem that has typically found you, a problem that left unresolved means trouble. When faced with a practical problem whose solution is not immediately obvious, you usually ask yourself a question whose answer you hope will help you solve the problem. But to find that answer, you must pose and solve a problem of another kind, a research problem defined by what it is that you do not know or understand, but feel you must. The process looks like this:

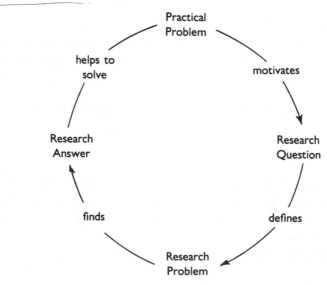

PRACTICAL PROBLEM: My brakes have started screeching.
RESEARCH QUESTION: How can I get them fixed right away?
RESEARCH PROBLEM: I need to find a nearby garage in the Yellow Pages.
RESEARCH ANSWER: The Car Shoppe, 1401 East 55th St.
APPLICATION TO PRACTICAL PROBLEM: Call to see when they can fix them.

It's a pattern common in every part of our lives:

- I want to impress a potential employer. *How do I find a good restaurant?* Look in a city guide. *Woodlawn Tap.* I take her there, and I hope she thinks I've got style.
- The National Rifle Association presses me to oppose gun control. *Will I lose if I don't?* Take a poll. *My constituents support gun control.* Now decide whether to reject the NRA's request.
- Costs are up at the Omaha plant. *What has changed?* Compare personnel before and after. *More turnover now.* If we improve training and morale, our workers stay with us. OK, let's see if we can afford to do it.

We don't write up solutions to most such problems, but we usually have to when we want to convince others that we have solved a problem important to *them:*

> To CEO: Costs are up in Omaha, because your workers see no future in their jobs and after a few months quit. You have to train new ones, which is costly. To retain workers, upgrade their skills so they will want to stay.

Before anyone could solve the *practical* problem of rising costs, though, someone had to solve a *research* problem defined by not knowing why costs were rising.

4.1.2 Distinguishing Practical Problems and Research Problems

This distinction between practical, pragmatic problems and research problems may seem to be a fine one, but it is crucial:

- A *practical* problem originates in the world and exacts a cost in money, time, happiness, etc. You solve a practical problem by changing something out there in the world, by *doing* something.

But before you can solve a *practical* problem, you may have to pose and solve a *research* problem.

- A *research* problem originates in your mind, out of incomplete knowledge or flawed understanding. You might pose a research problem because you have to solve a practical problem, but you do not solve that practical problem merely by solving the research problem. You might *apply* the solution of that research problem to the solution of a practical problem, but you solve your research problem not by changing anything in the world but by learning more about something or understanding it better.

Most medical researchers, for example, believe that before they can solve the practical problem of the AIDS epidemic, they must first solve in the laboratory a research problem posed by the puzzling mechanism of its virus. But even if medical researchers solve that research problem by discovering its mechanism, governments still have to find a way to apply that solution to the practical problem of AIDS in society.

"Problem" thus has a special meaning in the world of research, one that sometimes confuses beginning researchers who usually think of problems as "bad." Every researcher needs a "good" research problem to work on; in fact, if you don't have a good research problem, you have a practical problem that is bad indeed.

4.1.3 Distinguishing Problems and Topics

There is a second reason that beginning and even intermediate researchers struggle with this notion of "problem." Experienced researchers often talk about their research problem in a shorthand way that seems to describe it just as a topic: *I'm working on adult measles,* or on *early Aztec pots,* or on *the mating calls of Wyoming elk.*

As a result, many beginning researchers confuse having a *topic* to read up on with having a *research problem* to solve. Lacking the focus provided by the search for a solution to a well-defined research problem, they just keep gathering more and more data, not knowing when to stop. Then they struggle to find a principle for deciding what to include in their report and what not, and finally just throw in everything they have. Then they feel frustrated when a reader says, *I don't see the point; this is just a data dump.*

You risk wasting your reader's time if you cannot distinguish between a *topic* to read about and a research *problem* to solve. In the rest of this chapter, we explain what a problem is, both academic and nonacademic. We return to problems in Chapter 15, when we discuss how to state your research problem in the introduction of your paper.

4.2 THE COMMON STRUCTURE OF PROBLEMS

We have distinguished pragmatic problems and research problems, but they have the same essential structure. Both consist of
two elements:

(1) some particular situation or condition, and
(2) its undesirable consequences, *costs* that you don't want to
pay.

4.2.1 Practical Problems

A flat tire is usually a practical problem, because it is (1) a
condition out there in the world that (2) may exact from you a
tangible cost—perhaps missing a dinner engagement. But suppose
your dinner companion bullied you into accepting the date and you
would rather be anywhere else. In that case, the flat tire does not
have a cost, because now you judge missing that dinner date to be
a positive benefit. In fact, the flat tire is now not part of a problem,
but part of a solution.

So when you think you have found a problem, be sure that
you can identify and describe a situation with these two parts:

- a *condition* that needs to be resolved
  CONDITION: I missed the bus.
             The hole in the ozone layer is growing.
- *costs* of that condition that you don't want to bear
  COST: I may lose my job because I will be late for
        work.
        Many will die from skin cancer.

You can often rephrase negative costs in positive form, as a benefit
of resolving the condition:

BENEFIT: If I can catch the bus, I save my job.
         If we fix the ozone hole, we save many lives.

The greater the consequences of the condition—either the costs
of leaving it unresolved or the benefits of resolving it—the more
significant the problem.

For a practical, tangible problem, the condition can be literally
anything, even a seeming stroke of luck, if it has a cost: *You win the
lottery.* That might not seem like a problem, but what if you owed

a loan shark $5,000,000 and your name gets in the paper? Winning the lottery could then cost you more than you won: someone finds you, takes your money, and breaks your leg.

4.2.2 Research Problems

A practical problem and a research problem have the same structure, but they differ in two important ways.

Conditions. While the condition part of a practical problem can be any state of affairs, the condition of a research problem is *always* defined by a rather narrow range of concepts. It is always some version of your *not knowing* or *not understanding* something that you think that you and your readers should know or understand better.

That's why in Chapter 3 we emphasized the value of questions. Good questions are the first step to defining your research problem, because questions imply what you and your readers don't know or understand but should: *What role does genetics play in cancer? How do icebergs influence the weather? How did Latin epics influence Old English poetry? How much does the death penalty deter murder?*

Costs. The second difference is harder to grasp. It is that the consequences of a research problem might have nothing immediately to do with the world. The *immediate* cost or benefit of a research problem is always some *further* ignorance and misunderstanding that is *more* significant, *more* consequential than the ignorance or misunderstanding that defined the condition.

This idea of cost is easy to understand in a practical problem because its costs are usually palpable—pain and suffering, lost money, opportunity, happiness, reputation, and so on. The costs of a research problem, though, are that we do not know or understand something else. That's why the problem of a visit from the loan shark seems easier to grasp than the problem of not understanding the influence of Latin on Old English poetry. The costs of the first are more palpable than those of the second. But not understanding the influence of Latin on Old English poetry has costs nonetheless. If we do not understand those influences, we will not understand *something yet more significant*—what an important but puzzling poem might mean, what Old English poets knew and didn't know about other literatures, why Old English poetry is the way it is.

An advanced researcher must show that because she does not know or understand one thing, she cannot know or understand something else *more important*. She must answer the question, *So what?*

> *So what if I never understand the role of genetics in cancer, why cats rub their jaws against us, how bridges were built in ancient Greece? If I never find out, what greater cost do I pay in my larger knowledge or understanding?*

In short, you have no research problem until you know the cost of your incomplete knowledge or flawed understanding, a cost that you define in terms of a yet greater ignorance or misunderstanding.

4.2.3 When a Research Problem Is Motivated by a Practical Problem

It is easier to identify costs and benefits of a research problem when it is motivated by a practical problem:

> *So what if we don't know why costs are up in Omaha? We go bankrupt.*
> *So what if we do not understand the role of genetics in cancer? Until we do, we will not know whether we can identify the genes that predispose us to cancer, when it can be predicted, or even cured.*

The cost of not knowing the role of genetics in cancer is that we do not understand its cause. Or putting this in the form of a benefit, perhaps only when we understand the genetics of cancer can we cure it. Now we instantly recognize the additional costs of our ignorance and the benefits if we remedy it, because a solution to the research problem points to a solution to the practical problem.

But how can stories about the Alamo or the aesthetics of Tibetan weaving be part of a significant research problem? We see a condition clearly enough: incomplete knowledge. But what costs do we bear if we go on knowing incompletely?

> *So what if we don't know about the evolution of medieval plumbing or the life cycle of a rare orchid in central New Guinea? What's the cost if we never find out? Or the benefit if we do? Well, let me think . . .*

It is at this point that researchers invoke the idea of "pure research" as opposed to "applied research."

> ### Practical vs. Research Problems:
> ### A Typical Beginner's Mistake
>
> A practical problem with its tangible conditions and costs is easier for beginning researchers to understand and more interesting to study, so they are often tempted to take on as their topic a tangible problem in the world—abortion, acid rain, homelessness. That's fine, as a starting point. But you risk a mistake if you make a problem in the world the problem you try to solve in your research. No research paper can solve the problem of acid rain, but good research might give us knowledge that could help us solve it. Research problems involve only *what we don't know or fully understand*. So write your paper not to solve the problem of acid rain, but to solve the problem that *there is something about acid rain that we don't know or understand*, something that we need to know before we can deal with it. ⸸

4.2.4 Distinguishing "Pure" and "Applied" Research

In much academic writing, we don't try to explain the cost of our ignorance by showing how our research will improve the world. Rather, we show how, by not knowing or understanding one thing, we and our readers cannot understand some *larger and more important matter that we have an interest in understanding* better. When the solution to a research problem has no apparent application to a practical problem, but only to the scholarly interests of a community of researchers, we call that research "pure" as opposed to "applied."

For example, none of your three authors knows how many stars are in the sky (or how much "dark matter"), and, candidly, we don't feel bad about not knowing. We wouldn't mind knowing, but we can't think of any cost if we never find out, or any benefit if we do. And so for us, not knowing is no problem.

But for astronomers, *their* not knowing that number is part of a "pure" research problem of great significance *to them*. Until they know that quantity, they can't calculate another that is much more important—the total mass of the universe. If they could calculate the mass of the universe, they might discover something *more important still:* whether it will keep expanding until it peters out into oblivion, collapse back on itself to explode again into a new universe, or settle into an eternally steady state. Knowing the number of stars in the sky may not help solve any tangible problem in the world, but for those astronomers (and maybe some theologians), that number represents a gap in their knowledge that exacts a great cost: it keeps them from understanding something more significant—the future of the universe. (Of course, if you have an interest in knowing whether the universe has a future, then perhaps you can see how not knowing how many stars are in the sky could be part of a problem for you as well.)

You can tell whether a research problem is pure or applied by looking at the last of the three steps in defining your project:

Pure Research Problem:

1. *Topic:* I am studying the density of light and other electromagnetic radiation in a small section of the universe,

2. *Question:* because I want to find out how many stars are in the sky,

3. *Rationale:* in order to understand whether the universe will expand forever or contract into a new Big Bang.

This is a *research* problem because its question (step 2) implies that we do not know something. This is a *pure* research problem because its rationale (step 3) implies not something that we will do, but something we do not know but should.

In an *applied* research problem, the question still implies something we want to know, but the rationale in step 3 implies something we want or need to *do:*

Applied Research Problem:

1. *Topic:* I am studying the difference between readings from the Hubble telescope in orbit above the atmosphere and readings for the same stars from the best earthbound telescopes,

2. *Question:* because I want to find out how much the atmosphere distorts measurements of light and other electromagnetic radiation,

3. *Rationale:* in order to measure more accurately the density of light and other electromagnetic radiation in a small section of the universe.

4.2.5 Is Your Problem Pure or Applied?

You distinguish between a pure and applied research problem by the consequences you name in the statement of its rationale (step 3). In pure research, the consequences are conceptual and the rationale defines what you want to *know;* in applied research, the consequences are tangible and the rationale defines what you want to *do.*

Perhaps one of the biggest reasons beginners have a hard time getting the hang of pure research is that its costs are entirely conceptual, and so it seems to them less like curing cancer and more like counting stars. Feeling that their findings aren't good for much, they try to cobble the solution of a research problem onto the solution of a practical problem:

> If we can understand how politicians used stories about the Alamo to shape opinion in the nineteenth century, we could protect ourselves from unscrupulous politicians and be better voters today.

1. *Topic:* I am studying the differences among various nineteenth-century versions of the story of the Alamo.
2. *Question:* because I want to find out how politicians used stories of great events to shape public opinion,
3. *Rationale:* in order to help people protect themselves from unscrupulous politicians and become better voters.

In some areas this is a respectable strategy, some would say a preferable one. But in our example, the writer is unlikely to convince many readers that his research on the Alamo stories can in fact improve democracy.

In order to formulate an effective applied research problem, you have to show that the rationale named in step 3 is plausibly connected to the question named in step 2. You can test this by working back from the rationale. Ask yourself this question:

(a) *If my readers want to achieve the goal of* [state your objective from Step 3],

(b) *would they think that the way to do that would be to find out* [state your question here from Step 2]?

The more strongly your readers would answer "yes" to your question, the more effectively you have formulated the applied problem.

Try this test on the applied astronomy problem:

(a) *If my readers want to* measure more accurately the density of electromagnetic radiation in a section of the universe,

(b) *would they think that the way to do that would be to find out*
how much the atmosphere distorts measurements of it?
Since astronomers have decades worth of data collected from high-
powered telescopes on earth, their answer would seem to be *Yes:*
if they can discover how much the atmosphere distorts readings,
they could adjust all of their data accordingly.

Now try the test on the Alamo problem:

(a) *If my readers want to achieve the goal of* helping people protect
themselves from unscrupulous politicians and be better voters,
(b) *would they think a good way to do that would be to find out* how
nineteenth-century politicians used stories of great events to
shape public opinion?

In this case, readers would have a harder time seeing a connection between the goal and the research. A researcher who wanted
to help voters protect themselves might think of other courses of
action before he turned to nineteenth-century stories of the Alamo.

A reader might think that the following question defines a good
research problem, but one that is pure rather than applied:

1. *Topic:* I am studying differences among nineteenth-
   century versions of the story of the Alamo,
2. *Question:* because I want to find out how politicians
   used stories of great events to shape public opinion,
3. *Rationale:* in order to show how politicians use elements of popular culture to advance their political goals.

At the heart of most research in the humanities and much in
the sciences and social sciences are questions whose answers have
no direct application to daily life. In fact, in many traditional disciplines, researchers value pure research more than they value applied
research—as the word "pure" suggests. They see the pursuit of
knowledge "for its own sake" as reflecting humanity's highest calling—to know more and understand better, not for the sake of
money or power, but for the sake of the good that understanding
itself can bring.

If you pose a question of pure research as though you could
directly apply its answer to a practical problem, your readers may
think you naïve. When you pose such a question and you want to
discuss the tangible consequences of its answer, formulate your
problem as the pure research problem that it really is and then *add*
to that problem a further possible significance:

1. *Topic:* I am studying the differences among various nine-teenth-century versions of the story of the Alamo,
2. *Question:* because I want to find out how politicians used stories of great events to shape public opinion,
3. *Rationale:* in order to understand how politicians use elements of popular culture to advance their political goals,
4. *Significance:* so that we will know more about protecting ourselves from unscrupulous politicians and become better citizens.

If your project is more pure than applied but you still believe that it has indirect tangible consequences, you should say so. But when you state your problem in your introduction (see Chapter 15), formulate it as a pure research problem whose rationale is based on conceptual consequences; save the possible tangible consequences for your conclusion (see Quick Tip, pp. 252–53).

4.3 FINDING A RESEARCH PROBLEM

What distinguishes great researchers from the rest of us is the brilliance, the knack, or just the good luck of stumbling upon a problem whose solution makes everyone see the world in a new way. Fortunately, the rest of us can usually recognize a good problem when we bump into it, or it into us. As paradoxical as it may seem, though, most of us begin a research project not entirely certain of what our problem is, and sometimes just clarifying the problem will be our major result. Some of the best research papers do no more than pose an important new problem in search of a solution. Indeed, finding a new problem or even clarifying an old one is often a surer way to fame and (sometimes) fortune than solving a problem already there. So do not be discouraged if you cannot formulate your problem fully at the outset of your research. Remember, though, that thinking about it early can save you wasted hours along the way and especially toward the end.

Here are some ways you can aim at a problem from the start.

4.3.1 Ask for Help

Do what experienced researchers do when they are not clear about the problem they think they are investigating: talk to people. Talk to your teachers, relatives, friends, neighbors—anyone who might be interested in your topic and your question. Why would anyone need to answer your question? What would they do with an answer? What further questions might your answer raise?

If you are free to select your own topic, you might look for one that is part of a larger problem in your field. You will be unlikely to solve it, but if you can slice off a piece of it, your project will inherit some of its significance. (You will also be educating yourself about the problems of your field, no small dividend.) Ask your instructor what he is working on and ask to work on part of it.

A warning: If your teacher helps you define your problem *before* you begin your research and gives you leads on sources, do not let those suggestions define the limits of your effort. You must find other sources, bring something of your own to the definition of the problem. Nothing more dismays a teacher than a student who does exactly what was suggested, *and nothing more.*

4.3.2 Look for Problems as You Read

You can often find a research problem if you read critically. As you read a source, where do *you* feel contradictions, inconsistencies, incomplete explanations? Where do you wish a source had been more explicit, offered more information? If you are not satisfied with an explanation, if something seems odd, confused, or incomplete, tentatively assume that other readers would or should feel that way too. Experienced researchers have the confidence to assume that when they read a passage that they do not entirely understand, then something is wrong, not with them, but with what they are reading. In fact, when they cannot quite grasp something, they predictably assume that their source is wrong and that they may have found a new problem: an error, discrepancy, or inconsistency that they can correct.

Of course, you *may* be the one who is wrong, so if you make your disagreement the center of your project, re-read your source to be sure you understand it. The problem may have been resolved in a way that your source did not state. Research papers, published and unpublished, are full of useless refutations of a point never made in the first place.

Once you think you have found a real puzzle or error, try to do more than merely point it out. If a source says X and you think

Y, you have a research problem only if you can show that readers who go on believing X will misunderstand something more important yet.

Finally, read the last few pages of your sources closely. It is there that many researchers suggest more questions that need answers, more problems in search of a solution. The author of the following paragraph had just finished explaining how the daily life of the nineteenth-century Russian peasant influenced his military performance.

> And just as the soldier's peacetime experience influenced his battlefield performance, so must the experience of the officer corps have influenced theirs. Indeed, a few commentators after the Russo-Japanese War blamed the Russian defeat on habits acquired by officers in the course of their economic chores. *In any event, to appreciate the service habits of Tsarist officers in peace and war, we need a structural—if you will, an anthropological—analysis of the officer corps like that offered here for enlisted personnel* [our emphasis].

4.3.3 Look for Problems in What You Write

There is another way that critical reading can help you discover and formulate a good research problem: you can read your own early drafts *critically*. When you draft, you almost always do your best thinking close to the end, in the last few pages. It is then that you begin to formulate your final claim, which can often be turned into the solution to a research problem that you have not yet completely formulated.

When you finish your first draft (we may seem to be getting ahead of ourselves here, but we warned you that doing research was not a neatly linear process), you should look closely at your last two or three pages.

1. Look first for the main point of your paper, the sentence or two that would stand as your most important claim.

2. Next look for signs that your point has resolved a puzzle, settled conflicting opinions, revealed something not previously known.

3. Now try to ask a complicated question that your main point would plausibly answer. That question should define the

condition of ignorance or misunderstanding that, lacking your answer, you and your readers will continue to suffer.

When you can do that, you have defined the condition of your research problem, what you do not know but want to. The next step is easy: Ask *So what?* The harder step is answering. But if you can find an answer, you have successfully reasoned backward from your solution to a full statement of the problem you have solved (we return to this process in Chapter 15).

4.3.4 Use a Standard Problem

Every problem is different, but most problems fall into just a few categories, many defined by a researcher disagreeing or contradicting some generally held view. When you reach a point where you think you may have the outlines of a problem, look at the Quick Tip on "contradictions" after Chapter 8. You may recognize in that list a kind of problem you can work toward.

4.4 THE PROBLEM OF THE PROBLEM

Your teachers understand that you are not a professional, but they believe it important that you develop and practice the habits of mind of a serious researcher. They want to see you do more than just accumulate facts about a topic, then summarize and report them. They want you to formulate a problem that you (and perhaps even they) have a stake in seeing solved. You take your first step toward real research when you recognize a question that is significant to *you*, a question that you want to answer just for your own satisfaction, to satisfy your own desire to know more, to resolve a discrepancy, to settle a contradiction, regardless of whether anyone else cares. If you can do that much in your earliest research, if you can find some puzzle that you *care* about resolving, you have achieved something quite significant that will gratify your teachers.

Eventually, though, as you move on to advanced work, when you decide that you have reason to share your new knowledge and understanding with others, you will have to take this next step: You must try to understand what *your readers* consider interesting questions and problems, the costs *they* perceive resulting from a gap in *their* knowledge or flaw in *their* understanding. You take the biggest step of all when you not only know the kind of problem that your readers like to see solved, but can persuade them to entertain problems of a new kind. No one ever takes all three steps the first time out.

To work your way through all of this, you can use the three-steps we discussed in the last chapter. We change the language from *discover* to *show* and *understand* to *explain,* but the second and third steps still implicitly define your problem:

1. *Name your topic:*
   I am *writing about* _____,
2. *State your indirect question (and thereby define the condition of your problem):*
   . . . because I am trying to *show you* who/how/why _____
3. *State how your answer will help your reader understand something more important yet (and thereby define the cost of* not *knowing the answer):*
   . . . in order to *explain to you* how/why _____.

All this may seem disconnected from the real world, but it is not. Research problems in the world at large are structured *exactly* as they are in the academic world. In business and government, in law and medicine, no skill is more highly valued than the ability to recognize a problem important to a client, employer, or the public, and then to pose that problem in a way that convinces readers that the problem you have discovered is important to *them* and that you have found its solution. The work you are doing now is your best opportunity to prepare for the kind of work that you will have to do, at least if you hope to thrive in a world that depends not just on problem solving but on problem finding. To that end, no skill is more useful than the ability to recognize and articulate a problem clearly and concisely, an ability in some ways even more important than solving it. If you can do that in a class in medieval Chinese history, you can do it in a business or government office downtown.

CHAPTER FIVE

From Questions to Sources

If you are a beginning researcher and you don't know your library well, use this chapter to develop a plan for your research. If you are somewhat experienced, skip to the next chapter. If you are an experienced researcher, go to Part III.

ONCE YOU FORMULATE a few research questions, or even just a clear sense of a plausible topic, you can start hunting for sources. If you found your topic in an academic book or article, you have a trailhead: you can track down footnotes and bibliography and find other sources like it in the catalogue. But if you don't know where to find resources, you may feel you are facing a desert. It is a stressful moment when you want to look for information and don't know where to begin.

A riskier moment is when you know where the sources are, plunge in without a plan, and find yourself lost in a thicket of information. Sources can lead you anywhere and everywhere, so it is easy to lose yourself wandering from one lead to another. There's nothing wrong with aimless browsing, and much that is right. We three authors do it often. Everyone who loves learning finds time to wander through worlds of ideas. Indeed, that's how many important discoveries have been made, through serendipity—the chance encounter with a new problem or relationship no one could have foreseen. Examples range from penicillin to the adhesive that makes Post-It notes so useful.

Unfortunately, you can't rely on chance to produce good research. Faced with a deadline, you have to limit your browsing and develop a few good questions that focus your efforts. But focused questions do not come easily, and gathering more information is usually easier and always more entertaining than thinking through the value of what you have already found.

In short, if you have a deadline, you need a plan. In this chapter we'll talk about the resources you can look for and how to narrow them to a manageable list. In the next chapter, we will discuss how to work with your resources once you've found them. We will lay out this plan as if you could follow it step-by-step. In fact, you will probably navigate your search in ways that loop you back as much as they move you ahead.

5.1 FINDING INFORMATION IN LIBRARIES

You will find most of your resources in a nearby library. You may, of course, find that the only library near you lacks books and journals that your topic requires. Or you may find one that specializes in a particular historical period, such as the W. A. Clark Library in Los Angeles; in a cause, such as the National Rifle Association Library in Fairfax, Virginia; or even in a person, such as the Martin Luther King Library in Atlanta. But no matter how small, your library probably offers more help than you might suspect, including the following:

1. Reference librarians.

2. General encyclopedias and dictionaries, such as the *Encyclopaedia Britannica* and the *Dictionary of National Biography.*

3. General bibliographical guides, such as the *Readers' Guide to Periodical Literature.*

4. Card or computerized catalogues, including computerized bibliographies and databases.

In a larger library, these can direct you to specialized resources:

5. Specialized encyclopedias and dictionaries such as the *Encyclopedia of Philosophy* and the *Dictionary of Computing.*

6. Specialized bibliographies, abstracts of articles, books, and dissertations, reviews of the year's work in a particular field.

7. Guides that summarize the resources available for research in a field, where to find them, and how to use them.

5.1.1 Librarians

If you know your library, start looking for sources. If this is your first shot at serious research, you might first talk to a librarian. Large libraries even have librarians who specialize in particular topics. They are usually eager to help when you don't even know where to start, much less where to go next. If you feel too shy or proud to ask, get over it. Talk to your librarian. Others do it all the time.

As we have stressed, though, the most important work you do is *planning*. You will save days of work if you prepare specific questions (and you will avoid wasting your librarian's time). If you are not prepared, no librarian can help you. Early on, before you have a focused problem, your questions may be general: *Which guides to periodicals list articles about educational policy in the fifties?* But as you narrow your topic, try to ask questions that will help your librarian understand exactly what you need to know: *How do I find court decisions on the separate-but-equal doctrine in educational policy in the early fifties?*

> One new graduate student at the University of Chicago needed three trips to find where the university's research library keeps most of its books. She spent her first two trips wandering through the seven floors of Reading Rooms, finding only reference works. Only on the third day did she work up enough nerve to ask a librarian, who pointed to a door that led into the stacks. Moral of the story: Ask.

5.1.2 General Reference Works

You will find two kinds of help in general reference works such as the *Encyclopaedia Brittanica* or more specialized ones such as the *Encyclopedia of Philosophy*. First, you can get a standard overview of your topic. Second, at the end of the article, you may find a list of sources that could be your entryway into the library's catalogue. If you find nothing, your topic may be categorized under a different heading. For example, the 1993 *Books in Print* listed nothing under *gender,* the term that many researchers in women's studies prefer to use, but it had many entries under *sex.*

5.1.3 The Library Catalogue, Card and Computerized

Now go to your catalogue, whether it's a card drawer or a computer terminal. Look up the titles that you found in the reference works. (Be aware that not all libraries list *all* their holdings on-line. Check the physical card catalogue for older works.) If you did not find any sources in those reference works, you will have to start fresh. Look under not just the first headings that occur to you but *all* the headings that are in any way related to your subject.

If you find a promising source in the catalogue, look at the subject headings; they will direct you to other books on your topic. If it's a card in a drawer, you can find the subject headings along the bottom of the card. If it's a computer screen, you'll have to look around since different systems have different interfaces. But

> A quick way to expand on a small catalogue is to consult *Books in Print*. It lists by subject and author the books for sale in a given year. Your library may have editions from prior years. If you have enough time, libraries can borrow books not in their collections.

somewhere in the entry for your source you'll see a list of subject heads or "keywords." Your source is also catalogued under those headings, which means that they may lead you to other books related to your topic. If you have found a single recent book on your topic, look at the back of its title page: you will find there subject headings for more books on the same topic.

The listings in a large library may seem overwhelming. The University of Chicago library has 280 books on Napoleon, 2,826 books with the word "environment" in their titles. If the number is large, narrow the list, using the techniques we talked about in Chapter 3.

In a small library, you may find no promising titles on your first pass. When that happens, you must rely on your own ingenuity. Think of all the ways in which your topic could be described. If your library has a computerized catalogue, you can search for subject headings by typing in just one or two words. The computer will find sources with those words in their titles and subtitles. Once you find a book that seems useful, the computer will show you on a "next page" screen its pertinent bibliographical information.

If you exhaust the terms you can think of and still find nothing, you could be onto an important question that nobody has thought about before, or at least not for a long time. Centuries ago, for example, the subject of "Friendship" was important to philosophers, but was later ignored by the major encyclopedias. Recently, though, it has reemerged as a major topic. On the other hand, if you find nothing, your topic may be too narrow or too far off the beaten track to yield quick results. In either case, chances are you'll make something of your topic only through your own hard thinking. In the long run, it might make you famous, but it is not a topic for a paper with a close deadline.

5.1.4 Research Guides

Every major field has at least one guide to the resources that experienced researchers commonly use: lists of bibliographies, locations of important primary materials, research methods, and so on. If you aspire to become a professional in a field, you must spend time with such guides, particularly if your library holds materials that the guides cite. The first step in learning the ropes of research is to learn where the ropes are stored.

5.1.5 Specialized Bibliographies

You should be able to find at least one annual bibliography covering either your whole field or a specific aspect of it. If you are lucky, you will find an *annotated* bibliography that focuses on an area close to your problem. In addition to listing books and articles on a subject, it briefly describes them. In fact, an annual annotated bibliography can be the best way to get a quick overview of what other researchers think. Most fields also have a scholarly journal that reviews new research annually, which is even more useful.

The *Chronicle of Higher Education* lists new books monthly, and many journals list "books received" (books that publishers send hoping the journal will review them). Such lists are the most current bibliographical sources.

A final note: In the last few years, the technology for storing and retrieving information has improved dramatically. In some areas, CD's store bibliographies on thousands of articles, monographs, and other publications. While those resources are not available in every library, large ones have them in abundance. Ask the librarian to show you how to use whatever electronic databases are available.

5.2 GATHERING INFORMATION FROM PEOPLE

Most projects can be done from books alone, but you may also need information that is available only from people.

5.2.1 Experts as Sources of Bibliography

At every stage of research, you can usually find someone to guide you. At first, your teachers will help you focus your question and start collecting information. Here, too, the quality of the help you get will depend on the quality of the questions you ask. The more you *think* before talking to your teachers, the better you can explain what you are doing and the more helpful they can be. Your teachers may not have all the answers, and you will have to look for help from others. (You might even hope your teachers *don't* have the answers, because you will have something to teach them, and they will read your report with greater interest.)

You can never decide in advance how much help of this kind you will need. At one extreme, we know of a graduate student who met with his adviser every day for breakfast, reporting what he had found the day before and receiving guidance for the day before him. (It's a good thing for students that they rarely get that much help from anyone.) At the other

---

Three Kinds of Sources

PRIMARY SOURCES: These are the materials that you are directly writing about, the "raw materials" of your research. In fields that study writers or documents, the texts you write about are primary sources. In fields such as English or history, you usually cannot write a research paper without using primary sources.

SECONDARY SOURCES: These are books and articles in which other researchers report the results of their research based on primary data or sources. You quote or cite them to support your own research. If a researcher quoted your paper to support his argument, your paper would be his secondary source. If, on the other hand, he were writing your biography, your paper would be a primary source.

TERTIARY SOURCES: These are books and articles based on secondary sources, on the research of others. Tertiary sources synthesize and explain research in a field for a popular audience or simply restate what others have said. Tertiary sources can be helpful in the early stages of your research, but they make weak support for your argument because they often oversimplify and overgeneralize, are seldom up to date, and are usually distrusted by experts.

---

extreme are those fiercely independent scholars who disappear into the library and never talk with anyone until they emerge with their project completed. (We don't actually *know* any, but we assume they must exist somewhere.) Most researchers choose a middle way, relying on casual conversations to guide their reading, which stimulates more questions and hunches to try out on others.

A new source of bibliographical assistance is the electronic "bulletin board" or "list" available through the internet, the computer network known as "the information highway." The system has discussion groups on almost every conceivable interest, some very specialized. In addition to lists on such topics as educational testing, cognitive psychology, and the history of rhetoric, there are also lists for goat fanciers, Missouri spelunkers, and Morris dancers. So if you want to research goats, there's someone out there who can help you.

You get access through your computer office or by finding a teacher in your area of interest who is "on the net." A common "posting" is a request for bibliographical references. A recent posting on a list for historians asked about the origin of footnotes. The questioner was referred to an article written by the person responding! Of course, this source of assistance is inappropriate for beginning researchers, but if you are an advanced student stuck for a reference to an obscure topic, there is certainly someone in some interest group who can help you.

5.2.2 People as Primary Sources

In some areas, you may have to collect primary data from people. We cannot explain the complexities of interviewing, but you should remember one similarity between learning from people and learning from books: the more you sort out what you know from what you want to know, the more efficiently you will find what you need. In short, plan. Not that you have to script an interview around a set list of questions—in fact, that's a bad idea—but do prepare so that you don't use people aimlessly. You can always go back to a book, but people are not sources that you can return to repeatedly because you did not prepare well enough to get what you needed the first time.

Even if your research is not directly about individuals, you may still find people willing to provide information, if you can help them understand your interest in what they know. Don't ignore people in local industrial, governmental, or civic organizations. For instance, in addition to reading court cases concerning the separate-but-equal doctrine that your reference librarian helps you locate, you might also call the local school district to see whether anyone there has memories she would be willing to share.

5.3 BIBLIOGRAPHICAL TRAILS

As you can see, research is never solitary. Even when you seem to work alone, you walk in the footsteps of others, profiting from their work, their principles and practices. One fundamental principle is that you share the basis of your research by documenting your sources so that others can follow you, a practice whose value you will appreciate when you get down to work. Once you locate only one or two sources on a topic, you are onto a trail of research that can lead you wherever you need to go.

In a book, skim the preface. It may list the author's friends and family, but also those who the author thinks have done good work. Next, skim the bibliography and index. The bibliography lists books and articles on the same or related topics, and the index will show which were used most often (generally, the more pages devoted to an author or book, the more important they have been for the writer). Articles usually begin with an account of previous research, and most have footnotes or a list of references.

Now comes the second round. If your list is short, read everything on it. If it is long and you need to shorten it, start with sources mentioned by most of the works you read in the first round. As you proceed, focus on works most relevant to your problem. Do not, however, ignore a work that is not mentioned but is on your topic—you will get credit for originality if you turn up a good source that few others have found. By following this bibliographic trail, you can find your way through even the most difficult research territory because one source always leads to another.

> Caution. If you find a book that seems crucial to your work, be sure it is the most recent *edition* of that work. You can check whether there is a later one by consulting the *Library of Congress Catalogue.*

5.4 WHAT YOU FIND

Once you consolidate your leads, you should have a substantial list to guide the first stage of your reading. If you can afford to, buy important books or photocopy important passages. You can save hours of note-taking if you own a work and can legitimately highlight passages that you think you will use. (We need not dwell on the fact that marking up a library book violates a first principle of every research community: preserve sources for those who follow. If you must keep notes in a book, insert sheets of paper between pages or use big Post-it notes that you can remove.) You will profit even more if you get in the habit of summarizing what you have read *in writing*. The more you write along the way, the more easily you will face that looming first draft.

Among these resources you will probably find titles right on your question. You may even experience that moment of panic when you discover *your* title: "Transforming the Alamo Legend: History in the Service of Politics." At that moment, you might think, *There goes my project, nothing new for me to say.* You might be right, but probably not. Study the source to see if it settles *your* question. If it does, then you have to formulate a new one. But when you see how your topic has been treated by another, you will probably find something new to say. In fact, with the help of someone who has worked through your topic before you, you can usually ask a better question. Or it may be that the author has gotten things not quite right. If so, you have found an unwitting friend.

CHAPTER SIX

Using Sources

IF YOU CAN GATHER INFORMATION and report it accurately and intelligibly, you have a skill valued highly in both the classroom and workplace. More valuable yet is the ability to work through conflicting opinions and arguments, to weigh data of different kinds and from different sources, to bring together information not usually conjoined, and to arrive at an original slant on an important problem. To do that, you need to learn how to analyze your sources not just accurately but critically.

6.1 USING SECONDARY SOURCES

Many published reports are useless, even harmful, because their authors substituted speedy note-taking for thoughtful, critical reading. Here are the first two principles for using sources: One *good* source is worth more than a score of mediocre ones, and one *accurate* summary of a good source is sometimes worth more than the source itself.

Those principles seem obvious, but evaluating sources is a difficult art. Ask anyone taken in by con artists in print: *I thought it was true because I saw it in Reader's Digest*—the sad words of those who discover too late how easily dishonest or careless "researchers" can make bogus results seem plausible and get them published. *Nine out of ten doctors agree* . . . Well, which doctors? Polled when and how? Behind every "miracal cure" there is a "study" that "proves" its superiority over its rivals, but many do not stand up under close scrutiny.

> One of Booth's students got a summer job doing "scientific research" for a drug company. He was assigned to go through stacks of doctors' questionnaires and shred enough of them until nine out of ten of those left did indeed recommend the company's product. The preserved bogus files "proved" the case. The student quit in disgust, quickly replaced, no doubt, by someone less ethically careful.

When research is distorted, though, it is usually inadvertent. Fraud does occur, but research published by respected journals is almost always done by those who would never deliberately misrepresent their results. Yet ask almost any scholar whose work has been discussed by others, and she will tell you her work is as often as not reported inaccurately.

Sometimes misreporting happens when a lazy researcher relies on hearsay. Colomb heard a prominent researcher confess after her talk that she had never read an author whose work she had just discussed. Booth has been "refuted" by a critic who apparently read only a section title, "Novels Must Be Realistic," and did not know that in attacking the title he was agreeing with Booth's argument. Sometimes reports are both misquoted and misunderstood. One reviewer misquoted Williams and then, thinking he was disagreeing with him, used the misquoted evidence to argue for the point Williams originally made.

More distortions, though, result from conviction that grows too passionate: some researchers become so committed to their case that they find support for it wherever they look. They don't quite "cook their evidence," but they reach too far for their proofs. And of course there is always simple human error: a word dropped, a quotation mark omitted or ignored.

6.2 READ CRITICALLY

How do you deal with sources that might be unreliable, and how do you avoid misreporting them? Here are a few suggestions helpful to any beginner, perhaps even to established scholars.

6.2.1 Evaluate Your Sources

1. Take seriously our advice about narrowing your sources to the few most valuable to *your* inquiry. In the early stages, this means a lot of skimming of books and articles to identify which ones you want to know better. Of course, you will make mistakes as you practice this speedy, and in some sense careless, reading. And you will have to re-read carefully. But only by skimming a lot can you settle on a few sources that deserve your most careful attention.

2. Once you locate a source that seems crucial, read *all* of it. In contrast to speedy reading, you must now read *slowly* to get a sense of the whole argument in its complete context. A common cause of misunderstanding is piecemeal reading—what is called "raiding." If you expect to use an argument or an idea, especially

if you intend to quote it, read everything around it and anything else that you need to understand what you expect to use.

3. If you use primary data or a quotation that you find in a secondary source, attribute that material to the primary source but acknowledge as well the secondary source in which you found it. More important, if your source relies significantly on an earlier source, check that source too. If you cannot find the quoted source, so be it; but if you can track it down, do so. You will soon discover that you cannot trust researchers to quote reliably. It is intellectually lazy not to look up an important quotation in its original context if that source is easily obtainable.

6.3 TAKE FULL NOTES

You can easily lose what you gain by careful reading if your notes do not reflect the quality of your thinking. Some believe that the best notes are kept on cards like this:

Sharman, <u>Swearing</u>, p. 133.   HISTORY/ECONOMICS (GENDER?)

Says swearing became economic issue in 18th c. Cites <u>Gentleman's Magazine</u>, July 1751 (no page reference) woman sentenced to ten day's hard labor because couldn't pay one shilling fine for profanity.

". . . one rigid economist practically entertained the notion of adding to the national resources by preaching a crusade against the opulent class of swearers."

(way to think about swearing today as economic issue? Comedians more popular if they use bad language? Movies more realistic? A gender issue here? Were 18th c. men fined as often as women?)

GT3080/S6

- At the top left of the card is the author, the title, and a page number.
- At the top right are key words that will let the researcher sort and re-sort cards into different categories.
- The body of the card contains a summary of the source, a direct quotation, and a thought about further research.

- In the lower right corner is the library call number for the book.

This format encourages systematic note-taking, but we three authors confess that we rarely use such cards. We record notes on a lined pad or on a computer, because a 3 × 5 space is too small for everything we want to say.

We should also point out that if you mix on the same card summary, paraphrases, quotations, and notes about your own thinking, you risk confusing them when you draft. It is safer to transcribe direct quotations, paraphrases, and close summaries on a card of one color, your own thinking on a card of another color, then paper-clip or staple the two cards together.

6.3.1 Get *Complete* Bibliographical Data

However you decide to take your notes, be certain that they record all the information you need to recover your critical reading and to let your readers know *exactly* how to find that same information. Here are some key elements.

Before you begin reading a work, record *all* of its bibliographical information. We can promise you that no habit will serve you better for the rest of your career. Record

- author,
- editor(s) (if any),
- volume,
- publisher,
- title (including subtitle),
- edition,
- place published,
- date.
- if an article is in an anthology or journal, all page numbers.

If you photocopy a section from a book, copy the title page as well and then write in the publication date from the reverse side of that title page. Finally, record the library call number of the book or journal. You won't cite this in your report, but most researchers can tell you how frustrating it is to find in their notes the perfect quote or the essential bit of data whose source was incompletely documented or not even identified. The call number will save you steps when you have to go back to the library to recheck a source.

If your source came over the internet, save all information about where and when you got it, not only the sender and date, but also the electronic source—a discussion or news list, a commercial database, etc. Many electronic sources are as public as libraries, but if you intend to quote something posted to a discussion or news list, it is a good idea to ask the sender's permission.

6.3.2 Get Attributions Right

When taking notes, you must clearly and consistently distinguish summary from paraphrase, and paraphrase from direct quotation. Make sure you put quotation marks around direct quotations and *avoid close paraphrases* (see pp. 166–71). Some researchers have seen their careers ruined when they published research that included a passage that they thought summarized what they had read, or even that they thought they had come up with on their own, and the passage turned out to be a direct quotation or a too-close paraphrase from a secondary source. When that was discovered, they were publicly accused of plagiarism. Their defense—in their notes they had inadvertently omitted quotation marks—may have been true, but it did them little good in the eyes of their research community. The best way to be certain that you distinguish the language of your source from your own and that your quotations are correct is to photocopy quotations longer than a few lines. Always record page numbers, not only of quotations and data, but of anything you paraphrase.

> A few years ago, Williams had to withhold publication of an article on Elizabethan social structure for a time because he failed to document a source fully. He had earlier come across information that no one else had thought to use for the problem he was addressing, but he could not use the data because he had failed to record complete information on the source. He searched the library at the University of Chicago for hours, until one night he sat up in bed, realizing the source was in a different library.

6.3.3 Get the Context Right

To support their claims, your sources build complex arguments out of several elements (we discuss this in detail in Part III). As you read your sources to assemble material for your own arguments, you should be analyzing theirs.

1. When you quote or summarize a source, be careful about context. You cannot completely avoid quoting out of context, because you obviously cannot quote all of an original. But if you read carefully, and re-read everything crucial to your own conclusions, your summaries and quotations will be made within the context that matters most, *the context of your own grasp of the original*. When you use a claim or argument, look for the *line of reasoning* that the author was pursuing and note it:

> NOT: "Bartolli (p. 123): The war was caused by Z."
> NOT: "Bartolli (p. 123): The war was caused by X, Y, and
>   Z."
> BUT: "Bartolli: The war was caused by X, Y, and Z (p. 123).
>   But the most important cause was Z (p. 123), for three
>   reasons: Reason 1 (pp. 124–26); Reason 2 (p. 126); Rea-
>   son 3 (pp. 127–28)."

Sometimes you will care only about the conclusion, but experienced researchers never just add up votes—*Four out of five sources said X, so I do too.* Readers want to know which conclusions result from *arguments,* those of your sources and especially your own. When you take notes, record not only conclusions but also the chief arguments that support them. That way, you'll be working in the context of *argued and related points.* (See Part III.)

2. When you record the claims made by your source, note the relative rhetorical importance of that claim in the original: Is it a main point? a minor point of support? a qualification or concession? a framing suggestion not a part of the main argument? Avoid this kind of mistake:

> Original by Jones: "We cannot conclude that one event causes
> another just because the second follows the first. And statistical
> correlation can never prove causation. But nobody who has stud-
> ied the data doubts that smoking is a causal factor in lung
> cancer."
>
> Misleading Report about Jones: "Jones makes the point that 'we
> cannot conclude that one event causes another just because the
> second follows the first. And statistical correlation can never
> prove causation.' No wonder responsible researchers distrust sta-
> tistical evidence of health risks."

Jones did not make this point at all. He merely *conceded* a point that he stated was relatively trivial as compared to what he said in the final sentence, which is the point he really wanted to make. Anyone who deliberately misreports in this way violates basic standards of truth. But a writer might make such a mistake inadvertently if her notes record only the words without noting their role as a minor concession.

Be especially attentive to "framing" statements at the beginning and end of an argument. Even careful scholars frame their discussions with large contextualizing statements. Sometimes those are their most interesting claims, but while they may believe them, they do not always try to support them.

Distinguish statements that are central to an argument from qualifications or concessions the author acknowledges but downplays. Unless you are reading a source "against the grain" of the writer's plan—for example, you want to expose hidden tendencies—do not report minor aspects of a research report as though they were major ones, or worse, the only information in the report.

3. Be sure about the scope and level of confidence an author expresses in making a claim. These are not the same:

X seems often to cause Y.

X causes Y.

4. Don't mistake the summary of another writer's views for those of the author summarizing them. Many writers do not clearly indicate through a long report that they are summarizing another's arguments, so it is easy to quote those authors as saying the opposite of what they in fact believe.

5. When dealing with sources that agree on a major claim, decide whether they also agree on how they interpret and support that claim. For example, two social scientists might claim that some social problems are caused not by environmental forces but by personal factors, but one might support that claim with evidence from genetic inheritance while the other points to religious beliefs. How and why sources agree is as important as the fact that they do.

6. When dealing with sources that disagree, be sure to locate the source of the disagreement. You need to know whether they disagree over evidence, over their interpretation of the same evidence, or over their basic approach to the problem.

Do not attach yourself to what any one researcher says about your subject. It is not "research" if you simply summarize and uncritically accept another's work. If you rely on at least two sources, you will almost always find that they do not agree entirely, and that's where your own research begins. *Which has the better argument? Which better respects the evidence? Is there an even better account that subsumes or refutes one or both of them?* In short, at this stage be critical of your sources; guard against being easily convinced by any of them.

Finally, remember that your report can be accurate only if you double-check your notes against your sources. After your first draft, check your quotations against your notes. If you use one source extensively, skim its relevant parts after you have finished your draft. By this time, you may be in the grip of the enthusiasm we mentioned earlier. You'll *believe* in your argument so strongly that you will see all your evidence in its favor. Despite our best intentions, that temptation afflicts us all. There is no cure, save for checking and rechecking. And rechecking again.

6.4 GET HELP

As your research progresses, you face a growing danger that you will collect information faster than you can handle it. Most researchers face that confusing moment when everything they have learned runs together. While they know a lot, they can't be sure what's really useful. You can't expect to

> Whether you are a beginner or expert, mistakes are part of the game; all three of us have discovered them in things we have published (and hoped that no one would find them). Mistakes are most likely when you copy a long quotation. When Booth was in graduate school, his bibliography class was told to copy a poem *exactly* as written. Not one student in his class of 20 turned in a perfect copy. His instructor said that he had given that assignment to hundreds of students, and perfect copies had been returned by just three. So check everything more carefully than you think necessary. But do not feel that you are the only one ever to make an especially silly mistake. Booth still winces when he remembers the graduate paper he turned in on Shakespeare's *McBeth.* And Williams would prefer to forget the report he was supposed to give but never did because he could find no references to his assigned topic, that great Norwegian playwright, Henry Gibson.

avoid all such moments, but you can minimize the anxiety they create by taking every opportunity to organize and summarize what you have gathered *in writing* and *as you go*.

At such moments, you can again turn to friends, classmates, teachers—anyone who can be a sympathetic but critical audience. Pause regularly to explain what you have learned to nonexperts. Try to articulate a coherent account of how and why what you have learned bears on your question and moves you toward a resolution of your problem. Give your friends progress reports and then ask them questions: *Does this make sense to you? Am I missing an important aspect or question? Given what I have said, what else would you like to know?* You will profit from their reactions, but even more from the mere act of explaining your ideas to nonspecialists.

At first you may find it awkward to ask others to listen to your ideas, but don't let that stop you. Make a deal with some classmates that you will help them if they will help you. Form a study group with three or four people who will listen to one another report on their work. Researchers do this all the time. The three of us would never submit a research report to a journal or to a press until we tried it out in public; and before that, we try out our ideas on friends, often on each other. In fact, this book grew out of such conversations, out of testing ideas over coffee.

Quick Tip:
Speedy Reading

You owe your readers a careful reading of your important sources to be certain that you report not only their main points reliably, but also their contexts, qualifications, and connections. But to discover which sources deserve a detailed reading, you must know how to do a speedier kind of reading to select out those works that are likely to be most important. Such speedy reading cannot be done just by running your eyes over the words of a source.

To identify quickly and reliably the main elements of an argument, you must know where to look for them. To do that, you must understand both the structure of an argument (a matter we discuss in Part III) and the geography of the book or article that reports it (the topic of Part IV). If you are ready to read your sources but have not yet read those two parts, do that first, then review this Quick Tip before you head off to the library.

As you read quickly, your goal is an overview of what your source offers: its topic, research problem, resolution, and the outlines of its argument. At this point, take only enough notes to remind yourself of its gist. You might then set that source aside, but it could turn out to be relevant later as your project develops.

STEP 1: Become Familiar with the Geography of the Source.

Before you begin skimming a source, get a sense of its whole.
1. If your source is a book,
   - read the first few sentences of each paragraph in the preface;
   - look at the Table of Contents for prologues, summary chapters, etc.;
   - skim the index for those topics with the most page references;
   - skim the bibliography, noting dates (current is best, of course) and sources cited most often;
   - flip through the chapters to see if they are divided

into sections with headings and if they have summaries at the end.

If your source is a very long book, a short published review of it
can give you a sense of its argument, major claims, and probably
an idea of its structure. (Look for a review in the relevant bibliographical source: see pp. 273–90.)

    2. If your source is an article,

- read the abstract, if it has one;
- flip through its pages to see if there are section
  headings;
- skim the bibliography.

STEP 2: LOCATE THE POINT OF THE ARGUMENT.

Read the introduction, particularly its last few paragraphs,
then the conclusion. In one or the other, you will find a statement of the problem and its resolution. Identify as well the kind
of evidence that supports the main claim.

STEP 3: IDENTIFY KEY SUB-POINTS.

Once you have some notion of the problem and its resolution, you can either reject your source as irrelevant or set it aside
for later close reading. If you cannot yet decide, look for the major sub-points that support the main claim.

    1. For a book or article, repeat step 2.

    2. If the chapter or article does not have headings, identify
its major chunks. Look for places where the writer wraps up
one major topic and introduces another with transitional words.
Train your eye to look for transitions ("First . . . Second . . .
Third . . ." "Finally," or "Now we have to consider Y").

    3. In each chunk, read the first and last paragraph, looking
for its major claim. Try to identify the kind of evidence used in
the chunk.

STEP 4: IDENTIFY KEY THEMES.

Once you have notes on the problem, main claim, and supporting points, scan the source for key concepts. List those concepts along with any bibliographical information on your source.
That list of words will help when you review your notes to see

whether sources you did not read carefully at first might later seem worth a closer look.

STEP 5: (IF NECESSARY) SKIM PARAGRAPHS.

Steps 1–4 will likely give you the information you need to decide whether to read a source more carefully, but if you are still unsure, skim each paragraph, looking for its point or main idea. If you find nothing in the first sentence or two that feels like a point, skip to the last one.

Whenever these five steps suggest that the source is relevant to your question, put it aside for more careful reading, a process that will be easier because you have already a sense of the most important features of its argument. As you will see when we turn to the matter of planning and doing your own first draft, practice in this kind of speedy reading can help guide your own strategy of writing and revision. If your readers cannot skim *your* reports and discover the outlines of your argument, the organization of your own writing will not have served them well.

PART THREE

Making a Claim and Supporting It

Prologue: Arguments, Drafting, and Conversations

FIRST THOUGHTS ABOUT A FIRST DRAFT

If you have accumulated a bushel of notes, photocopies, and summaries, and they are spilling off your desk or filling up your hard disk, it's time to think about a first draft. You may have only dim outlines of answers to your most important questions—in fact, you may not yet know exactly what they are. But once you accumulate a substantial body of data, you have to start thinking about what they add up to. One way to get closer to an answer is to sort through your materials in a way that will help you discover in them some pattern or implication and formulate a claim you think you could support.

When beginning researchers start to organize their material, too many sort it under the most obvious topics, arrange those topics into a plausible sequence, and start writing. Unfortunately, the most obvious topics may be the least useful, because they probably reflect not what you discovered through your own hard thinking, but only what your sources gave you. And even if those topics do go beyond the obvious, they are likely to constitute only a linear sequence (A + B + C + . . .), a rhetorical structure rarely strong enough to support a long and complex argument. The worst result is that you just summarize someone else's ideas.

To be sure, sorting is a good way to *prepare* for a first draft— sort your data by any topics that seem appropriate. Finally, though, as you close in on the moment when you have to start planning a first draft, you need a principle of organization that comes not from the categories of your data but from your questions and their answers. You have to organize those answers to support some central *claim* that you want to make, a claim that will stand as the answer to your hardest question, as justification for writing your paper. The support for that answer and claim will take the form of a research *argument*.

ARGUMENT AS CONVERSATION

In Chapter 4, we distinguished everyday, troublesome problems from the kind that motivate research projects. In the same way, here we have to distinguish between everyday arguments and the kind that organize research reports. People usually think of arguments as disputes: children argue over a toy; roommates over the stereo; drivers about who had the right-of-way. Such arguments can be polite, but they always imply conflict, with winners and losers. To be sure, researchers sometimes wrangle over evidence, jockey for position, and occasionally erupt into charges of carelessness, incompetence, even fraud. But that is not the kind of argument that made them researchers in the first place.

In the next four chapters, we examine a kind of argument that is less like a contentious debate and more like a thoughtful conversation in which, together with others, you explore ideas on issues that you all believe are important.

In that conversation, however, you do more than just exchange views. We are all entitled to our opinions, and in everyday conversation no law requires that we explain why we hold them. But in the world of research you are expected to make claims that you think are new and important enough to interest your readers, and then you are expected to explain those claims as if your readers were asking you, quite reasonably, why you believe them. Because you anticipate those questions, you support your claims with good reasons and grounds, with *evidence*.

You should also know, however, that readers you respect will question your evidence, perhaps even your logic, and so you must explain your argument as well, breaking it into subordinate claims, themselves supported with further evidence. You may even feel that you must explain why you think your particular evidence logically supports your particular claim. Finally, you have to anticipate that readers will think of objections and alternatives, and so you have to answer them as they are likely to arise.

Your objective in all this is not to force your opinions down

the throats of your readers or to overwhelm them with unqualified
Truth, but, by anticipating their views, their positions, their interests, to put forward your claims in a way that helps them recognize
their own best interests. By helping you explore the limits of your
evidence and test the soundness of your reasoning, the elements of
a good argument help you work not against your readers but *with*
them to find and understand a truth you can share.

CHAPTER SEVEN

Making Good Arguments:

An Overview

In this chapter we survey the four elements of a research argument. Chapter 8 discusses the two elements essential to every argument, Chapters 9 and 10 two additional elements that advanced researchers must master and beginners should at least understand.

7.1 CONVERSATIONS AND ARGUMENTS

THERE IS NOTHING ESPECIALLY DIFFICULT about the kind of argument that you find or have to make in a research report. It has the same give-and-take of a lively discussion with those whose intelligence you respect, especially when their questions can help you think through to the solution of a complicated problem. The only difference is that in conversation you usually feel more confident about what you know, and another person is right there in front of you, asking questions that encourage you to think hard about what you believe and why you believe it:

A: So how do you think you'll do this semester? [*A asks a question, implicitly raising a problem.*]

B: I think I'll do better than last. [*In answering the question, B makes a claim and implicitly solves the problem.*]

A: Why do you think so? [*A asks for evidence to support the claim.*]

B: I'm finally taking courses in my major. [*B offers evidence.*]

A: Why will that make a difference? [*A doesn't see why taking courses in a major counts as relevant evidence.*]

B: I do better when I take courses that interest me. [*B offers a principle about courses and motivation that connects the claim to the evidence.*]

A: But what about that statistics course you have to take? [*A points to evidence that might counterbalance B's evidence.*]

B: I know I bombed calculus, but statistics is easier and now I have a tutor who can explain things better than my teachers do. [*B acknowledges the contrary evidence, but rebuts it by offering more evidence.*]

A: But won't you be taking five courses? [*A raises another reservation.*]

B: I know. It won't be easy. [*B concedes a point he cannot rebut.*]

A: Think you'll make the Dean's list? [*A asks about the limits of B's claim.*]

B: I can't promise, but I think I'll do well. At least a 3.00, as long as I don't have to get a part-time job. [*B limits the scope of the claim and then stipulates a condition that qualifies his confidence.*]

If you can imagine being a part of that conversation, you will find nothing strange in research arguments, because the elements are the same. The only difference is that in a research report, not only must you answer your readers' questions, you must also ask questions on their behalf. Their questions will include these:

Reader's Questions	Your Answers
What is your point?	*I claim that . . .*
What evidence do you have?	*I offer as evidence . . .*
Why do you think your evidence supports your claim?	*I offer this general principle . . .*
But how about these reservations?	*I can answer them. First, . . .*
Are you entirely sure?	*Only if . . . and as long as . . .*
No reservations here at all?	*I must concede that . . .*
Then just how strong is your claim?	*I limit it . . .*

Your answers constitute your argument. It should offer

- *a claim,*
- *evidence* or *grounds* that support it,
- something we call a *warrant,* a general principle that explains why you think your evidence is relevant to your claim,
- *qualifications* that make your claim and evidence more precise.

As you put arguments together, no habit of mind will serve you better than imagining yourself in a conversation with your readers, you making claims, your readers asking good questions, you answering them as best you can.

7.2 CLAIMS AND EVIDENCE

The two elements that you must always state explicitly are your claim and supporting evidence.

- your *claim* states what you want readers to believe;
- your *evidence* or *grounds* are the reasons they should believe it.

> *Claim:* It must have rained last night,
>
> *Evidence:* because the streets are wet.
>
> *Claim:* You should be checked for diabetes,
>
> *Evidence:* because your glucometer reading is 200.
>
> *Claim:* The emancipation of Russian peasants was largely symbolic,
>
> *Evidence:* because it didn't improve the quality of their daily lives.

When you offer either of these elements without the other, you seem to offer either pointless data or ungrounded opinion.

Claims and evidence are enough for ordinary conversations, such as whether it rained last night. But when you make a significant claim, you ask your readers to change their minds about something important. Since most readers correctly resist changing their minds easily, especially about important matters, you will usually need to expand your argument with two more elements: warrants and qualifications.

7.3 WARRANTS

The warrant of an argument is its general principle, an assumption or premise that bridges the claim and its supporting evidence, connecting them into a logically related pair. Your warrant answers questions not about whether your evidence is accurate but about whether it is relevant to your claim; or, to put it the other way around, whether your claim can be inferred from your evidence. Think of your warrant as a superstructure that bridges evidence and claim:

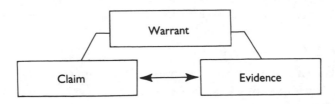

In casual conversation, we rarely ask for a warrant. If we asserted, *It must have rained last night because the streets are wet,* few would ask in response, *Why should the fact that these streets are wet make me believe your claim that it rained last night?* Almost everyone just takes for granted the warrant, the general principle, that links the evidence of wet streets to a claim about rain.

> *Whenever we see the evidence of wet streets in the morning, we can conclude that it probably rained the night before.*

(Of course, if you live in a village that uses sprinklers to keep the dust down, the warrant alone would not be enough; you would also want to know whether the sprinklers were operating last night. More about this in Chapter 9.)

For other kinds of claims, though, questions about warrants are inevitable. Suppose you get your blood tested at one of those booths set up in shopping malls. The volunteer reads the device that tests for blood sugar and says, *You should be checked by your doctor*$_{claim}$ because *your reading is 200.*$_{evidence}$ Almost all of us would ask why 200 means that we should see a doctor. When we do, we are asking for a warrant, a principle that justifies connecting particular evidence— 200 on this device—to a particular claim—that we should see a doctor. *Well,* responds the tester, *whenever someone has a reading of more than 120, that's a good sign that she may have diabetes.*

You often need to include the additional supporting structure supplied by an explicit warrant, because research arguments normally ask readers to change their minds about things that are not obvious. That often means you have to convince your readers that your evidence is in fact relevant to your claim.

For example,

> The emancipation of Russian peasants was merely symbolic,$_{,claim}$ because it didn't improve the quality of their daily lives.$_{evidence}$

This argument might lead a reader to ask for a warrant:

> *Even if I grant your evidence that the quality of life for Russian peasants did not improve, why should that lead me to believe your claim that their emancipation was merely symbolic?*

The researcher would have to respond with a general principle that states how a kind of evidence is relevant to a particular claim:

Whenever a political action fails to improve the lives of those it is alleged to help, we judge that reform to have been only symbolic.

Of course, the reader might reject that warrant as false. If so, he would then have to question the argument as a whole, *even though both the evidence and the claim might be factually true.* (We'll discuss all of this in more detail in the next two chapters.)

7.4 QUALIFICATIONS

The fourth part of an argument consists of *qualifications.* Qualifications limit the certainty of your conclusions, stipulate the conditions in which your claim holds, address your readers' potential objections, and—when not overdone—make you appear a judicious, cautious, thoughtful writer.

Whenever you make a claim that is true only under certain conditions or you assert a connection between evidence and claim that is not 100% certain but only probably true, you owe it to yourself and your readers to qualify your argument appropriately. When you qualify your argument, you acknowledge the obstacles that interrupt the movement between your evidence and claims.

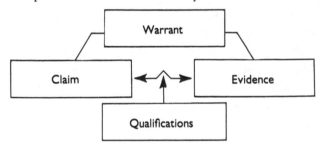

For example, a reading of 200 is not always a sign of diabetes. First thing in the morning, 200 is high, *unless* you just ate a big sweet roll. So before we can evaluate a claim and its evidence, we have to know how their scope is qualified: *Your reading is 200*qualification, *so you should be checked,*claim *because that much glucose in the blood is a* **good**qualification *sign that you* **may**qualification *have diabetes,*warrant **unless, of course, you just ate something sugary.**qualification

The more complex and interesting your argument, the more qualifications you are likely to need, because complex and interesting claims are never cut-and-dried, 100% true under all circumstances. To be sure, some great thinkers (and more than a few teachers) deliver Olympian judgments as if they were beyond qualification. The rest of us do better to qualify. Without "waffling" or "dodging the question," we should show legitimate caution about our results. (See pp. 141–42)

How you manage claims, evidence, warrants, and qualifications is important in how readers judge not just your arguments but the quality of your mind, even your character. Most readers will want to know why you make a claim, not to challenge you, but because they want to understand your argument better, to be part of the conversation. When you acknowledge their interest, you present yourself as a considerate writer. If you simply claimed, *You should be checked for diabetes* or *The emancipation of the Russian peasants was merely symbolic,* and said nothing more, you would seem to expect your audience to believe whatever you said simply because you said it, always an uncivil assumption. Good reasons and thoughtful qualifications help convince your readers that you are trustworthy.

When you make a claim, give good reasons, and add qualifications, you acknowledge your readers' desire to work with you in developing and testing new ideas. In this light, the best kind of argument is not verbal coercion but an act of cooperation and respect. But this structure of argument is more even than that. It can also guide your research. If you understand how your sources put together their arguments, you can read them more critically and take notes more accurately. If you understand how you will have to put together your argument, you can plan your first draft more efficiently and test your findings more reliably.

CHAPTER EIGHT

Claims and Evidence

In this chapter, we discuss the two elements that you must make explicit in every argument. These matters are important for everyone who wants to make a credible argument, beginning and advanced researchers alike.

THE CENTRAL ELEMENT IN EVERY REPORT is its major claim, its main point or general thesis. It is the culmination of your analysis, the statement of what your research means. But if you want your readers to change their minds about something important to them, you cannot simply assert that claim; you have to give them good reasons, reliable evidence for believing it. This pair, claim and evidence, constitutes the conceptual core of every research report.

8.1 MAKING STRONG CLAIMS

Your main claim is the heart of your report, the part that most fully reflects your personal contribution to your research. To hold up your end of the dialogue, that claim must meet the expectations of your readers. They expect it (as well as the subordinate claims that support it) to be *substantive, contestable,* and *explicit.*

8.1.1 Your Claim Must Be Substantive

Readers want you to help them understand something important, so they will take little interest in a claim only about what you have done:

> Though the 1981–82 recession occurred chiefly because OPEC raised oil prices, I have examined the role of the Federal Reserve Board.

or about what your paper will do:

> This paper will discuss the role of OPEC and the Federal Reserve Board in the recession of 1981–82.

Neither says anything substantive about OPEC, the Federal Reserve, or the recession, so neither needs an argument to support it. Such claims typically introduce an aimless walk through data.

The following claim might be substantive enough to engage a reader's interest, because it makes a claim about OPEC, oil prices, the Federal Reserve, the money supply, and the 1981–82 recession:

> The 1981–82 recession did not occur because OPEC raised oil prices but mainly because the Federal Reserve Board tightened the money supply.

8.1.2 Your Claim Must Be Contestable

Readers think a claim significant to the degree that it is contestable. It should lead them to think, *You'll have to explain that,* either because they have long thought otherwise, or because they never thought about it at all. No one contests a claim that refers only to the report itself or to you, nor a claim that repeats what readers already believe:

> Thus World War II changed the course of history by allowing the Soviet Union to dominate Eastern Europe for almost half a century.

Since most readers already believe this, saying so adds nothing new. If nothing you tell them changes their minds in ways they care about, you waste their time. Only to the degree that your claim changes what they already believe will it be contestable. To the degree it is contestable, your readers will think it significant. (See the Quick Tip at the end of this chapter for some common ways researchers make contestable claims.) But again, if this is one of your first research projects, focus on your own interests, on what would be significant to you, or to someone with *your* interests and knowledge.

8.1.3 Your Claim Must Be Specific

Readers also expect your claims to be couched in language sufficiently detailed and specific for them to recognize the central concepts that you will develop throughout your paper. Compare:

> Thus the emancipation of Russian peasants was not a significant event.

Thus the emancipation of Russian peasants was not significant, because while their lives changed somewhat, their situation declined.

Thus the emancipation of the Russian peasants was only symbolic, because while they gained control over their daily affairs, their economic condition deteriorated so sharply that their new social status did not affect the material quality of their existence.

The first claim has little substance. The second is less vague, but it announces few specific concepts that readers should watch for (except *decline*). The third is explicit, announcing several concepts that the author must develop in its support: *symbolic, gain control, economic condition, deteriorate, new social status, material quality of existence.*

When you state your main claim early, at the end of your introduction (as most readers prefer; see pp. 202–6), it is important that you state it in language that is specific. When readers see that language reappear, they are more likely to feel that your text is coherent. When readers do not know what concepts to expect, they may miss important ones and judge what they read to be unfocused, even an incoherent mess.

8.2 Using Plausible Claims to Guide Your Research

Your readers will dismiss your claims if they are not substantive, contestable, and explicit. Those qualities can also be important to you while researching and drafting. You will understand your sources better when you can identify their major claims and the evidence they offer in support. You give yourself directions for research when you create substantive claims with explicit topics and concepts: what would you need to flesh out *gain control, economic condition, deteriorate, new social status, material quality of their lives?*

You can also use these concepts to sort your evidence:

Before the peasants were emancipated, their **material lives** were sufficient for survival.
—*What evidence relates to "material lives"?*
Their social level was **low.**
—*What evidence relates to "low"?*
They did not **control** their lives.
—*What evidence relates to "control"?*

Their social status **rose** somewhat.
—*What evidence relates to "rising"?*
The material quality of their every day life **deteriorated.**
—*What evidence relates to "deteriorated"?*

Each term is simultaneously part of the main claim and of sub-arguments that will need their own supporting evidence. The more explicit your language, the more evidence you need to support your claims and the better you will see the research you must still do.

If you are writing your first research paper, the task of formulating a significant, contestable claim in richly specific language can seem impossible, especially when your reader is an expert in the subject of your research. *How,* you might ask, *am I supposed to find something that my teacher does not already know or believe?* Teachers understand this problem and will expect you to make a claim that would be new and contestable *for someone with your level of experience and knowledge,* perhaps just *new for yourself.* In that case, do your research with your own interests in mind, or those of your classmates. What might they find surprising, contestable, significant?

If, however, you are an advanced student, your teachers will expect you to make a claim that would be considered contestable—or at least worth testing—by experts. In that case, your research must include what experts now believe about your problem and how they have responded to similar ones. Ask your teacher what she expects.

8.3 OFFERING RELIABLE EVIDENCE

Your claim is the heart of your paper, but most of the paper will be devoted to supporting evidence. If readers reject your supporting evidence because they think it is weak, it will fail one or more of six tests: they will judge it not to be *accurate, precise, sufficient, representative, authoritative,* or *perspicuous.* (Readers may also reject evidence because it is irrelevant or inappropriate, but to test evidence by those two criteria you have to know more about warrants, which we discuss in the next chapter.)

These criteria are not unique to research arguments. We use them in our most mundane conversations. The following argument by C fails on all six criteria of quality, and on appropriateness as well:

C: I need new sneakers.*claim* These seem small.*evidence*

P: Your feet haven't grown that much in a month, and they don't seem to hurt you much. [*i.e., I accept that what you offer as evidence could be relevant to your claim, but I reject it first because it is not accurate and second because even if it were accurate, "seem small" is not sufficiently precise.*]

C: But they look awful. They're dirty. Look at these raggedy laces.*evidence*

P: Raggedy laces and dirt aren't reason enough to buy new sneakers. [*i.e., Your assertion may be factually correct and with more evidence might be worth considering, but shoelaces and dirt alone are not sufficient evidence of the terminal condition of your shoes.*]

C: Everybody thinks I should get new sneakers.*evidence*

P: Everybody's opinion doesn't matter. [*i.e., Even if it's true, other people's opinions are to me not authoritative.*]

C: Haven't you seen the way I have to walk?*implied evidence*

P: No. [*i.e., How you walk might qualify as evidence, but I have seen you and I don't see anything wrong. Your evidence is not perspicuous.*]

C: Look at how I limp.*evidence*

P: You were walking fine a minute ago. [*i.e., Your evidence is not representative.*]

C: You can afford to buy me new shoes.*implied evidence*

P: Forget it! [*i.e., I will not respond because your evidence is not appropriate.*]

If you can imagine yourself as P, you can test the quality of evidence in any research argument.

8.3.1 Accuracy

Above all, your evidence must be accurate; expert readers are contemptuous of error. Read again our cautions in Chapter 6 about taking notes that accurately reflect both the text and context of passages you cite. (See pp. 77–80.) If your paper depends on data collected in a lab or in the field, record your data completely and clearly, then double-check both before and *as* you write them up. Readers predisposed to be skeptical of your arguments, as all thoughtful readers should be, can seize on the smallest flaw in your data, on the most trivial mistake in a quotation or citation (even your spelling and punctuation), as a sign of your irredeemable unreliability. Getting the easy things right shows respect for your readers and is the best training for the hard things.

Since accuracy is crucial, one way to sort your evidence is by its reliability. What evidence are you certain of? What evidence do you wish were more reliable? You can use questionable evidence, *if you acknowledge its quality.* In fact, when you point to evidence that seems to support your claim and then reject it as unreliable, you show yourself to be cautious and self-critical.

8.3.2 Precision

Researchers want evidence that is not only accurate, but precise. What counts as precise, though, differs from field to field. A physicist measures the life of quarks in infinitesimal fractions of a second, so the tolerable margin of error is vanishingly small. A historian gauging when the Soviet Union was ready to collapse would estimate it in weeks or months. A paleontologist dating a new species would give or take tens of thousands of years. According to the standards of their fields, all three are appropriately precise. (Evidence can also be too precise. A historian would seem foolhardy if he asserted that the Soviet Union reached its point of collapse at 2 P.M. August 18, 1987.)

Though you should not make your evidence seem more precise than it is, careful readers will hear warning bells if you use certain words that so hedge your claim that they cannot assess its substance:

> The Forest Service has spent a **great deal** of money to prevent forest fires, but there is still **a high probability** of **large, costly** ones.

How much money is "a great deal"? How high is a "high" probability—30%? 50%? 80%? How many acres are destroyed in a "large" fire? Watch for words like *some, most, many, almost, often, usually, frequently, generally,* and so on. These qualifiers can set appropriate limits on a claim, but they can also fudge it. (We'll return to qualifications in Chapter 10.)

8.3.3 Sufficiency

Just as different fields judge the precision of evidence differently, so they differ in gauging its sufficiency. In some fields, researchers base a claim on evidence from a single episode of research: a critic calls a new novel a potboiler after one reading and cites as evidence a single flaw. For a claim about handedness and baldness a psychologist might want results from 150 subjects in a dozen experiments. But before accepting a new cancer drug, the FDA would demand data from thousands of subjects through years of trials. The more at stake, the higher the threshold of sufficiency. It might be interesting to know whether a new novel is a potboiler or more left-handers are bald, but few suffer from wrong results. Not so with a new cancer drug.

Beginners typically present insufficient evidence. They think they have proved a general claim when they find support in one quotation, one bit of data, one personal experience:

> Shakespeare must have hated women because in *Macbeth* they are all either evil or weak.

Researchers almost always need more than one bit of data to support a claim that is substantive and contestable (though sometimes only one bit of evidence will disprove a claim). If you are making a claim even mildly contestable, offer your best evidence, but know that there is *always* more available, and that it could contain counterexamples fatal to your claim.

Paradoxically, some beginning researchers cite the very lack of evidence as proving their claim:

> No evidence shows life elsewhere in the universe, so there must be none.

You can see how useless negative evidence is when you recognize that, on the same question, it can cut both ways:

> No evidence shows that life cannot exist elsewhere in the universe, so it probably does.

8.3.4 Representativeness

Data are representative when their variety reflects the variety of the population from which they are drawn and about which you make your claim. What counts as representative also varies by field. Anthropologists might interpret a small culture in New Guinea on the basis of a deep acquaintance with a few individuals, but no sociologist would make a claim about American religious practices based on data from a single Baptist church in Oregon. Beginners always risk offering evidence that fails to reflect the range of available evidence, not because they are careless, but because they cannot imagine what more representative evidence would look like.

Once you collect your evidence, ask your teacher or someone experienced in the field about the kind of additional evidence they would expect to support a claim such as yours. If you want to learn how to judge this for yourself, ask your teacher for examples of arguments that failed because they were based on unrepresentative evidence. We learn what counts as representative by accumulating representative examples of what does not.

8.3.5 Authority

Competent researchers cite the most authoritative sources, but what counts as authoritative again varies by field. Note the authorities whom researchers in your field cite most often, what procedures they trust, what records they regularly cite. If you are dealing with primary sources (original texts of books, plays, diaries, and so on), be sure that your edition is recent and published by a reputable press. There are on-line electronic editions of Shakespeare so sloppily edited that using them would label you as incompetent.

When students are unfamiliar with or can't find authoritative secondary sources—scholarly journals and books—they often resort to tertiary sources: textbooks, articles in encyclopedias, mass-circulation publications like *Psychology Today* (see our warnings on p. 69). If these are the only sources available, so be it, but never assume that they are authoritative. Be especially cautious about books on complex issues intended for mass audiences. Authors who write for the ordinary reader about brains or black holes are usually competent, sometimes distinguished researchers. But they must always simplify, sometimes oversimplify, and are always out of date. So if you start your research with a popular book, look at the dates of the journals cited in its bibliography.

Authority also depends on being current, but again, different fields judge currency in different ways. In the sciences, out-of-date might be a month ago. In the humanities, a scholar might judge as reliable a book more than a century old. The best way to gauge currency is to skim dates in bibliographies of recent journal articles. What seems to be the cutoff date? Assume that most textbooks and reference books are out-of-date.

Remember, though, that some of the best research proves that a long-established "current and authoritative" notion is in fact not true. For decades people in many fields casually cited the "fact" that the Inuit peoples of the Arctic have dozens of terms for types of snow. Only when a researcher checked did she find that they in fact may have just three. (Or at least so *she* claims.)

Finally, distinguish authoritative evidence from "authorities." In every field, if Expert A says one thing, Expert B will assert the opposite. And someone else will claim to be Expert C, who in fact is no expert at all. When beginning researchers hear experts disagree, they (and the public as well) can become cynical about expertise and dismiss the experts' knowledge as mere opinion. Don't confuse uninformed cynicism with informed and thoughtful skepticism.

If you are an intermediate researcher, do not trust any source as authoritative until you know the research in the area. Nothing reveals incompetence more quickly than quoting someone whom everyone in the field scorns—or worse, has never heard of.

Different fields define all these criteria differently, but each demands that evidence meet them. Listen to lectures and class discussions for the kinds of arguments that your instructors criticize because they think the evidence is unreliable. Ask for examples of bad arguments, even if they have to invent them. You will understand what counts as reliable only after you see examples of what does not. Acquiring that knowledge through the mistakes of others is less painful than gaining it through your own.

8.3.6 Perspicuity

Your evidence may be accurate, precise, sufficient, representative, and authoritative, but if readers cannot *see* your evidence *as* evidence, you might as well offer no evidence at all. Especially when it consists of quantitative data or direct quotations, be sure that your readers can see in it what you want them to see. For example:

In the treadmill test, metabolic values for subjects 1, 3, 7, and 10 were invalid. Pulse rate data at 4, 8, and 10 minutes:

Subject	Rest	T = 4	T = 8	T = 10
1	61	72	93	101
2	73	88	105	110
3	66	85	99	110
4	73	88	105	110
5	66	85	99	110
6	81	97	111	124
7	81	97	111	124
8	73	88	105	110
9	66	85	99	110
10	81	97	111	124

What should we see here? We would know only if we already knew that metabolic effects occur when pulse rates per minute rise above 170% of the resting rate, and we can do the percentages in our head. Otherwise, these data feel not like evidence but like raw, undigested numbers. (In Chapter 12, we will offer some principles for analyzing and revising tables like this.)

Equally puzzling is the "bare bones" quotation. Here is a student's claim about Lincoln, citing as evidence the "Gettysburg Address":

> Lincoln believed that the Founders would have supported the North,$_{\text{claim}}$ because as he said, this country was "dedicated to the proposition that all men are created equal."$_{\text{evidence}}$

It may be that the Founders would have supported the North, but what in that quotation should make us think that Lincoln believed they would? When pressed, the writer explained:

> Since the Founders dedicated the country to the proposition that all men are created equal and Lincoln freed the slaves because he thought they were created equal, then he must have thought that he and the Founders agreed, so they would have supported the North. *It's obvious.*

Well, it's not. Quotations rarely speak for themselves; most have to be "unpacked." If you offer only evidence without interpretation, your report will seem to be a pastiche of quotations and numbers, suggesting that your data never passed through the critical analysis of a working mind.

Whenever you support a claim with numbers, charts, pictures, quotations—whatever looks like primary data—do not assume that what you see is what your readers will get. Spell out what you want them to see as the *point* of your evidence, its *significance*. For a quotation, a good principle is to use a few of its key words just before or after it. Introduce a chart, table, or graph by pointing out both what you want readers to notice and why it is notable.

To understand why evidence fails, you need experience and the ability to anticipate what readers are likely to accept or reject. You gain that ability in two ways. The more painful way is to be the object of criticism. Less painful is to hear from your teachers example of arguments that fail. By understanding failed examples, you can evaluate your own more objectively. So ask.

8.4 Using Evidence to Develop and Organize Your Paper

This scheme for evaluating arguments should encourage you not to approach your readers in a spirit of conflict or coercion. Rather than staking out a position and fiercely defending it against those whom you expect to attack it, imagine yourself and your readers in a civil conversation, working together to create new knowledge, the kind of conversation that you should have been having with *your* sources.

The emphasis on dialogue in this scheme can also help you find and build your arguments, especially when your notes seem nothing more than a pile of undigested information. As you prepare to draft, use the elements of argument as a principle of organization that helps you anticipate your readers' concerns. The scheme is useful even in the earliest stages of gathering information. If you understand how researchers put together their arguments, you can do a better job of reading your sources and taking notes on them.

As you review your data, remember that your argument must always be in the form of a claim plus supporting evidence. But you can't convince readers just by piling up data, because convincing reasons are not a matter of quantity alone, or even of their quality. Considerate researchers also (explain) their evidence. They present their evidence and then treat it as if it were a claim in a more detailed argument that requires still more evidence. As researchers construct supporting explanatory arguments to support their evidence, they offer readers good reasons why their evidence is sound.

In this next paragraph, the writer claims that the Forest Service has wasted millions, then offers evidence: despite all that money, there's been no decrease in the incidence of fires. But he doesn't stop there. He goes on to explain the evidence, pointing out that the total number of fires has remained constant but large fires have decreased. Then he explains why they have decreased.

> There is good reason to believe that since 1950, the U.S. Forest Service has wasted millions on trying to prevent fires when it could have better spent those resources on managing small ones that go out of control to cause catastrophic damage.$_{claim}$ Despite the millions spent on prevention, the number of fires in western forests has remained unchanged since 1930. But starting in 1950, the number of devastating fires began to drop, because it was then that the Service systematically used firefighting aircraft to reach small fires quickly and bring them under control before they could spread. Had the millions spent preventing fires since then been spent on efforts to keep the small ones from spreading, there would be fewer of those massive conflagrations whose costs dwarf the money spent on their prevention.

Every researcher must support contestable claims with evidence, but she must then explain that evidence, treating each major bit of evidence as a claim in a subordinate argument that needs its own evidence. In fact, every research report consists of multiple arguments of different kinds, but all in the service of the central claim that the researcher wants to make. So the structure of your paper will always be more elaborate (and less linear) than a single claim supported by a single body of evidence. The evidence that supports a main claim will itself be divided into groups of smaller arguments, each of them structured as a (sub-)claim with its own supporting evidence:

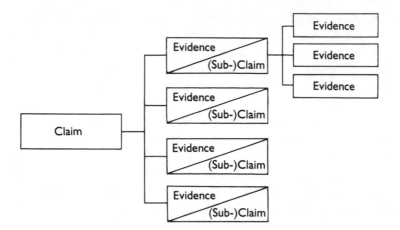

If you like doing things visually, put this on a wall-sized chart. Pin up index cards as in the figure above, then try different combinations of sub-arguments. Don't worry about organizing information within each card; just focus on getting it into middle-sized groups that you can arrange and then rearrange in a variety of configurations.

If this chart looks like an outline, it is. But it outlines not your paper but your argument. When you begin to outline your first draft, you'll have to think more about your readers: how to introduce your problem to make it seem significant to them, how much background to present, and how to order your subclaims, and so on. These are important matters, but they are not pressing when you are only at the point of discovering your argument.

This "Quick Tip" will probably be most useful to more advanced students, but beginners should become familiar with these kinds of contradictions because they will see them in everything they read.

You cannot determine how "significant" a claim is until you know how much others in your field must change their thinking in order to accept it. In all fields, though, a common way of implying significance is to contradict settled ideas. (And by claiming that something your readers believe is incomplete or incorrect, you create the condition of a problem. Review pp. 51–55.) We can't tell you what ideas you should contradict, but we can show you some standard *kinds* of contradictions that turn up again and again in research literature.

SUBSTANTIVE CONTRADICTIONS

If you can show that a previous researcher has gotten something wrong, you can easily signal the significance of your argument. The more authoritative the mistake, the greater the significance. Three cases are most common:
- You find an error in a fact or computation.
- You have new facts that either qualify old facts or replace them.
- You find a mistake in reasoning and from the same facts come to a different conclusion.

FEATURE CONTRADICTIONS

Other kinds of contradictions follow patterns that are so standard that they are like those categories of questions we encouraged you to ask about your topic (pp. 39–42). We do *not*, however, encourage you to memorize or to limit yourself to the items on this list. We offer them only as a way to encourage your thinking and imagination.

Category Contradictions

It has often been claimed that certain religious groups are "cults" because of how they differ from mainstream churches, but if we look at those organizations from a historical perspective, it is not clear when a so-called "cult" becomes a "sect" or even a "religion."

In this pattern, you claim that your argument contradicts the *categories* that others in your field accept. Generally, you promise to show either that while others place something in a category, they should not; that while others do not place something in a category, they should. (In the examples, substitute for X and Y terms of your own.)

    1. Though X seems to be an example of Y, it is not.

Though cigarettes seem to be addictive, they are not.

Or the case can be reversed:

Though cigarettes seem not to be addictive, they are.

Other common patterns of category contradictions:

    2. Though X seems to include Y as an example, it does not.
    3. Though X and Y seem to be similar, they are different.
    4. Though X seems to be characteristic of Y, it is not.

Part-Whole Contradictions

In recent years, some have argued that athletics is only entertainment and therefore should have no place in higher education, but in fact it can be shown that without athletics education would suffer.

This pattern is like the category contradiction, except that you show that others have mistaken the relationship among the parts of something.

    1. Though X seems to be an integral part of Y, it is not.
    2. Though X seems to have Y as an integral part, it does not.
    3. Though the parts of X seem to be systematic, they are not.
    4. Though X seems to be general, it is only local.

Internal Developmental Contradictions

Recently, the media have been headlining rising crime, but in fact the overall crime rate has been falling for the last few years.

In this pattern, you claim that others have mistaken the origin, development, or history of your object of study.

    1. Though X seems to be stable/rising/falling, it is not.
    2. Though X may seem to have originated in Y, it didn't.

3. Though both X and Y may seem to have come from Z, X didn't.
4. Though the sequence of development of X seem to be 1, 2, and 3, it is not.

External Cause-Effect Contradictions

A new way to stop juveniles from becoming criminals is the "boot camp" concept. But evidence suggests that it does little good.

In this pattern, you claim that others have either failed to see causal relationships or seen them where they do not exist.

1. Though X seems not to be causally related to Y, it is.
2. Though X seems to cause Y, both X and Y are caused by Z.
3. Though X and Y seem to correlate, they do not.
4. Though X seems to be sufficient to cause Y, it is not.
5. Though X seems to cause only Y, it also causes A, B, and C.

Value Contradictions

In this pattern, you simply contradict received valued judgments.

1. Though X seems to be good, it is not.
2. Though X seems to be useful for Y, it is not.

PERSPECTIVAL CONTRADICTIONS

Some contradictions run deeper. In the standard pattern of contradicting features, you reverse a widely held supposition, but you do not change the terms of the discussion. In perspectival contradictions, you step outside of the standard discussion to suggest that we must look at things in an entirely new way.

It has generally been assumed that advertising is best understood as a purely economic function, but in fact it has served as a laboratory for new art forms and styles.

1. We have generally discussed X in Y context, but there is a new context of understanding that we should consider— social, political, economic, intellectual, academic, gender specific, etc.

2. We have generally seen X as explained by theory Y, but there is a new foundational theory or a theory from another field that can be applied to X that makes us see it differently.
3. There is a new value system by which to evaluate X.
4. We long ago used to analyze X using theory/value system Y, then we rejected X as inapplicable to Y, but Y is relevant to X in a new way.

If you find a plausible contradiction of any of these kinds, keep track of it, because you will be using it when you write your introduction. More of that in Chapter 15.

Chapter Nine

Warrants

This chapter raises issues more complex than some beginning researchers might want to wrestle with. Advanced students, however, should ponder them.

GOOD RESEARCH SHOULD CHANGE OUR THINKING. It asks us to accept a new idea, or in the strongest case, to rearrange our system of beliefs in fundamental ways. Such changes we rightly resist without good reasons. So when you ask your readers to change their minds, you owe them your best reasons for doing so. But you can't just pile up more and more data, no matter how reliable, because good reasons go beyond their sheer quantity, even beyond their quality. Unlike those who never apologize and never explain, considerate researchers always ask themselves whether they need to explain why their data are not just reliable but *relevant*.

9.1 WARRANTS: THE BASIS OF OUR BELIEF AND REASONING

To explain why your data are relevant, you may have to articulate an element of your argument that is often left tacit. It shows readers why any particular body of data *should count* as evidence in support of your claim. This connection between claim and evidence is your *warrant*. Here again is the argument about wet streets and rain:

> *Claim:* It must have rained last night.
>> *Why do you think that? (That is, what's your evidence?)*
> *Evidence:* The streets are wet this morning.
>> *What makes you think that wet streets should count as evidence of rain? (That is, what's your warrant?)*

If we accept the evidence as reliable—that the streets were indeed wet this morning—what principle or premise, what underlying assumption must we accept before we believe the claim that it must have rained? It would be that wet streets *generally* mean rain, an assumption so obvious that we never bother to state it:

> *Warrant:* Whenever we see wet streets in the morning, we can usually conclude that it rained the night before.

A warrant is a *general principle* that creates a logical bridge between *particular* evidence (wet streets *this* morning) and a *particular* claim (it rained *last* night).

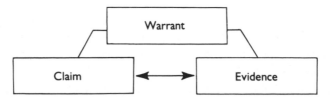

In the argument about wet streets, the connection is so obvious that you would never state it, nor would listeners expect you to. In fact, if you did, you might affront them by implying that they did not know a fact so obvious; and if they asked you to state your warrant, you would be just as affronted, for the same reason (unless you lived in a village that sprinkles its streets; we'll discuss such reservations in Chapter 10).

But when you construct complex arguments, especially ones that address contestable issues, your assumptions can betray you if you leave them unstated and unexamined. For example, here is a bit of an argument about the Forest Service that might give readers pause:

> *Claim:* The Forest Service has wasted money on fire prevention.
>
> *Evidence:* Since 1950, the Forest Service has spent millions preventing fires, but the number of fires has remained the same.

Warrants and Formal Logic

If you have had a course in formal logic, you may be wondering how warrants fit its categories. If you remember the term *major premise*, you can see that a warrant is analogous to the major premise in a *conditional syllogism* (If p,q; p; therefore q). But, as we shall see, a warrant also has features of a *categorical syllogism* (All B is C; A is B; therefore A is C). In this scheme, evidence roughly coincides with the minor premise.

The evidence is true. But why should it permit the writer to claim that money spent on fire prevention has been wasted? What else must we believe? It would be a warrant like this:

> *Warrant:* Whenever anyone spends money to prevent something but its incidence remains the same, that person has wasted the money.

At first glance, that warrant seems sound, but is it true *under all circumstances?* No exceptions? Have conditions changed—more tourists? drier climate? has the cost of prevention tripled?

Even experienced researchers can take their warrants too much for granted, because they are hidden in the theories that guide their research, in the definitions of their words, even in the metaphors they use. In this chapter, we will show you how not to take your own warrants too much for granted, how to decide whether a warrant is true, whether it in fact allows you to connect particular evidence to a particular claim, and when you should make warrants explicit. The concept of a warrant is difficult, but until you grasp it you risk building arguments that your readers might think are plainly illogical.

9.2 WHAT DOES A WARRANT LOOK LIKE?

When you state a warrant, you have to phrase it as a generalization that answers your reader's question, *What general principle must I believe before I can agree that your admittedly accurate evidence about these wet streets actually supports your at least plausible claim that it rained last night?* We can state such a warrant in many ways:

> Wet streets in the morning are a result of rain the night before.
> Rain at night usually means wet streets the next morning.
> A morning with wet streets is a sign of rain the night before.
> Rain in moon glow, wet streets at sun up.

But to qualify as a warrant, it must satisfy these three criteria:

1. • One part must describe the *general kind* of evidence you offer.
2. • The other part must describe the general kind of claim that follows from the evidence.
3. • It must state or imply a connection between them, such as cause-and-effect (Rain causes wet streets), one-

as-the-sign-of-another (Thunder is often a sign of rain),
many-instances-allow-a-generalization (A northwest
wind usually means a clear day).

(For other connections, see the Quick Tip at the end of this
chapter.)

But if warrants can be phrased in many ways, for the purpose
of our thinking about and analysing them one way is most useful:

When(ever) we have evidence *like* X, we can make a claim *like* Y.

In this scheme, you express in the first half of the warrant the
general *kind* of evidence or grounds that the warrant admits and in
its second half the *kind* of claim it allows. The logical connection
between the two is signaled by *when(ever)*. We can shorten all that:

When(ever) ~~we have evidence that~~ X [streets are wet in the
morning,] ~~we can usually claim that~~ Y [it probably rained the
night before].

leaving just, "When(ever) X, Y."

You may recall this way of formulating warrants from one of
the most important documents in American history:

. . . whenever any Form of Government becomes destructive of
[a people's right to life, liberty, and the pursuit of happiness], it
is the Right of the People to alter or to abolish it. . . . when a
long of train of abuses and usurpations [pursues] invariably the
[deprivation of those rights], it is [the people's] right, it is their
duty, to throw off such Government and to provide new Guards
for their future security.

When you write to an audience that shares your assumptions, you
rarely state your warrants so baldly. But when you write to those
who might not share your beliefs and might therefore reject your
evidence as irrelevant, you have to offer not just the evidence, but
explicit warrants, as well.

Perhaps that's why Thomas Jefferson stated his warrant not
once but twice. The *Declaration of Independence* challenged an older
warrant about the relationship between people and their government, so Jefferson may have decided that he had to make his new
warrant absolutely clear, particularly since he felt that "a decent

respect to the opinions of mankind requires that [we] should declare the causes which impel [us] to the separation." Had he left his logic implicit, he risked having the world wonder why he thought the colonists should throw off the rule of King George III just because he abused them. A Royalist, after all, might offer a competing warrant: *When a person is a king, he can do what he wants, so your list of King George's alleged offenses is irrelevant.*

Even if you do leave most of your warrants unstated, it is a useful exercise to articulate the most important ones, at least to yourself, so that you can test the conceptual foundations of your argument. Thinking about warrants helps you find the unreliable spots in your argument before your readers do. You may even have to defend your warrants, with an argument that supports them (or as Jefferson did, by appealing to a fundamental truth given directly to the human mind: "We hold these truths to be self-evident").

9.3 THE QUALITY OF WARRANTS

Readers resist claims for many reasons. Some reasons are unjustified: despite the truth of your argument, some readers are too fixed in their ways to change their minds, or they have interests that your claim threatens, or they just don't want to work hard enough to understand your exposition. On the other hand, readers justifiably reject claims that are poorly formulated or based on unreliable evidence. Even when your claim is clear and significant and your evidence reliable, they will still reject your argument if they think your warrant is false, unclear, the wrong kind for your research community, or if it does not validly admit your evidence.

These are not exotic criteria; we apply them in our most mundane conversations, even those between parents and children.

1. False warrant

   C: Everybody else has new shoes, I need some too.

   P: If everyone jumped off the cliff, would you? [*Your warrant is false if you assume that when everyone else gets something new you should too.*]

2. Unclear warrant

   C: Look at this ad.

   P: So? [*Even if what the ad says is true, I don't see what the ad has to do with me buying you shoes.*]

3. Inappropriate warrant

> C: You have enough money.
> P: Forget it! [*Your assumed principle—that whenever I can afford to buy you something, I should—is utterly inappropriate.*]

4. Inapplicable warrant

> C: You must not love me.
> P: Ridiculous. [*Your implied evidence is true: I am not buying you shoes. And let's grant that your warrant may be true—parents who do not love their children do not buy them shoes—but your claim is unwarranted, because it does not follow that when any particular parent does not buy kids shoes, that parent must not love them.*]

In each exchange, the evidence might be reliable: everybody else may have new shoes, the ad may make the shoes look good, P may have enough money, and, of course, P is not buying new shoes. But if you can understand why P still rejects each argument, then you can understand why, even when your evidence is reliable and your claims plausible, readers may reject your arguments if you bridge your evidence and claims by warrants that are false, unclear, inappropriate, or inapplicable.

If you are a thoughtful researcher, you will interrogate your argument at least once to be certain that its warrants link your evidence and claim reliably, an exercise that can make you rethink assumptions left unexamined for a long time, especially the fundamental assumptions of your field. That may open the door to still more research of the kind that leads to your most interesting and important results.

9.3.1 False Warrants

You test the truth of a warrant as you would the truth of any claim, because most warrants are just claims from higher-order, more general arguments, claims that need their own supporting evidence, just as (working *down* the argument chain) a bit of evidence is a claim in need of its own support.

What would be the warrant for the next little argument? In addition to believing the truth of the evidence, what else must we believe before we can accept it as support for the claim?

In the late 1930's Franklin D. Roosevelt could not have been a widely popular president$_{claim}$ because so many newspapers accused him of leading the country down the road to socialism.$_{evidence}$

As we said, although researchers articulate warrants in many ways, the most useful for examining them is to break the warrant into two explicit parts, one that expresses the general kind of evidence that the warrant admits and one that expresses the claim it allows:

> W_1: Whenever many in the mainstream press attack an American president for leading the country down the road to socialism,$_{evidence\ side}$ that president is not universally popular.$_{claim\ side}$

Once you state the warrant in this "evidence-thus-claim" form, you can test its power by formulating wider and narrower versions of it:

> W_2: Whenever any form of journalism attacks any leader for any reason in any way,$_{evidence\ side}$ that leader is not popular.$_{claim\ side}$
>
> W_3: Whenever Midwest Republican newspapers in the '30s attacked a president for leading America to socialism,$_{evidence\ side}$ he was unpopular with business interests.$_{claim\ side}$

What would lead us to accept any of these three warrants? It would be hard to accept the most general one (W_2), because we can think of so many counterexamples. We invite problems, however, when we draw the warrant too narrowly, as in W_3: if the evidence side of the warrant is virtually the same as the evidence offered to support the claim, then the argument is judged to "beg the question."

A good principle is to assume a warrant general enough to include at least one category larger than the evidence, but not so general that you open yourself to a myriad of exceptions: make "Roosevelt" not "any leader" but "a U.S. president," and make "newspapers" not "any form of journalism" but "mainstream press."

Be sure to test the truth of your warrant with words like "always," "everywhere," "invariably." When you see your argu-

ment in terms so strong, you recognize the qualifications you may
have to add and perhaps additional research you must do to support
your warrant. If you don't, some reader will. Testing the truth of
warrants is difficult, not just because people rarely think about
them. When you question your warrants, you question the conceptual foundations of your research community.

9.3.2 Unclear Warrants

Each research community has its own warrants, typically unspoken, hidden in its research procedures, even in its machines. Scientists who study the brain use as evidence "scans" from a magnetic
resonance imager, a device that pictures the brain's electrochemical
activity. When a researcher points to a red spot on a computer
screen and says, *This area is active when he visualizes absent objects,* she
draws a conclusion from a chain of arguments that are invisible to
outsiders.

When you take such warrants for granted, you too easily offer
evidence that *you* may think relates to your claim, but whose relevance may elude your readers. This often happens when you take
a shortcut through several connected arguments, skipping middle
steps. For example, if you are unfamiliar with some general truths
about English social history in the sixteenth century, this passage
will baffle you:

> In 1580, fewer than half the students at some colleges in Oxford
> University could legitimately sign their names "John Jones, Esq."
> or "Mr. Jones."*evidence* It would thus be more than 300 years before
> English universities would again be so egalitarian.*claim*

How do we get from sixteenth-century signatures to egalitarian
twentieth-century universities? By omitting intermediate steps:

> *In 1580, fewer than half the students at some colleges at Oxford Univer-
> sity could legitimately sign their names "John Jones, Esq." or "Mr.
> Jones."evidence*

> STEP 1: In late 16th-c. England, only a man in the relatively
> small class of men called "gentlemen" could legitimately sign
> his name with "Mr.," and only the son of a gentleman could
> sign with an "Esq."*warrant1* In 1580, fewer than half the students

at Oxford could legitimately sign their names with "Mr." or "Esq."$._{evidence1}$ Therefore, fewer than half the students in those colleges were gentlemen or their sons.$_{claim1}$

STEP 2: When social classes in a university population are generally proportional to their numbers in the whole population, a university can be judged egalitarian.$_{warrant2}$ The low number of university students in the late 16th c. who were gentlemen or their sons [from claim[1]] roughly reflects the fact that less than half the English population were then gentlemen or their sons [from "small class" in warrant[1]].$_{evidence2}$ Thus, these colleges were roughly egalitarian.$_{claim2}$

STEP 3: Repeat to show that between 1600 and 1900 more gentlemen than commoners attended Oxford, making it less egalitarian, but that after 1900 more commoners attended than gentlemen, making it once again more egalitarian.

It would thus be more than 300 years before English universities would again be so egalitarian.$_{claim}$

Only someone familiar with English history could understand how the evidence of sixteenth-century signatures could be relevant to a claim about twentieth-century universities. The rest of us are bewildered.

This kind of thing happens when beginners assume that a chain of connections obvious to them must be equally obvious to their readers, as did the student we cited in the last chapter who claimed,

Lincoln believed that the Founders would have supported the North,$_{claim}$ because as he said, this country was "dedicated to the proposition that all men are created equal."$_{evidence}$

Look closely at the steps of your argument to determine whether you have skipped some. If so, you may have to reconstruct them.

When testing your argument before you draft it, be explicit. But when drafting, you have to decide how explicit you can and should be. When you leave warrants implicit, you make an important social gesture. Members of a research community share countless warrants, because warrants constitute the fabric of common principles and unspoken truths that makes a community what it is.

When you assume those warrants, you assume membership in that community, both for you and for your readers. But as we've said, when you make warrants unnecessarily explicit, you may insult the readers you hope for most. As you gain experience and credibility, you show it not only by what you do say, but by what you don't have to (look again at the two examples about calcium blockers on pp. 12–13).

9.3.3 Inappropriate Warrants

Sometimes a warrant can be true for both you and your reader, yet the reader rejects your argument because its warrant is inappropriate to that reader's research practices. This usually happens when your warrants are appropriate in your own community but not in another. Since different research communities are defined partly by their different warrants, you cannot assume that a warrant accepted in yours will be accepted in another. And when readers reject your warrant as inappropriate, they will reject your evidence, not as false, but as strange or even bizarre.

For example, a student writing about Robert Frost's poem, "Stopping by Woods on a Snowy Evening," might plausibly argue:

> The sounds of the first stanza reinforce the idea of quiet, comforting woods, because most of the vowels are low/back and most of the consonants are soft and voiced:
> Whose woods these are I think I know.
> His house is in the village though;
> He will not see me stopping here
> To watch his woods fill up with snow.

The unspoken warrant is one that students of literature accept but seldom make explicit, because their community takes it for granted:

> When we hear soft dark sounds, we associate them with soft dark images.

But that *kind* of warrant is not among those assumed by researchers in other fields. A historian, for example, might want to claim that in the 1952 presidential election, voters liked Dwight

Eisenhower because they saw him as a father-figure. But she would be unlikely to construct an argument like this:

> The sound of the Eisenhower slogan, "I Like Ike," subliminally comforted voters. The sound of "I" is embraced in the sound of "Ike," and both nestle in the sound of "like," thereby doubly embracing the "I" in the father's comforting love.

A historian would scoff at any warrant such as,

> When the sound of one word occurs inside another, readers associate the meaning of the inside word with the outside word.

On the other hand, a psychologist might make this argument:

> In contrast to Adlai Stevenson's nasal twang, Eisenhower's deeper voice induced a sense of comfort. Among 78 subjects who listened to recordings of his voice for ten minutes, 56 had a decreased mean pulse rate of 3 beats per minute, lowered blood pressure of 3.6%, and lowered muscle tension of 7.9%.

The warrant here is something like,

> When measures of heartbeat, blood pressure, and muscle tension fall, a person is becoming more comfortable,

a warrant of the sort appropriate to the world of psychologists.

Laboratory evidence might be used to support the claim that the sounds of "Stopping by Woods" also make us comfortable, and such empirical evidence might appeal to certain psychologists. But while literary critics might accept the claim and evidence as individually plausible, they would scorn the argument, rejecting as utterly foolish any warrant that justifies measuring aesthetic response by attaching someone to a blood-pressure cuff.

The job of the beginning researcher is to understand which warrants go with which fields, something that comes only with experience. We understand that such advice must seem like saying, *You'll understand when you're older.* But this is one of those matters where only experience serves. You cannot know whether an argument will work until you know the warrants that your readers work with. That you learn only by living with them for a time.

9.3.4 Inapplicable Warrants

The last test for warrants addresses a matter that has vexed logicians for two thousand years: how does a warrant connect evidence to a claim *validly?* When evidence is unreliable, you can correct it; when it is unclear, you can clarify it. But when your argument is *unwarranted,* you have to fix it in a way that alters its *logical structure.* Even when your claim, evidence, and warrant are all true, your reader may still reject your argument as invalid if their relationship is unwarranted—and what counts in good research is not just the apparent truth of your conclusions but the quality of the reasoning that got you there.

Here is that simple example about rain again:

> It must have rained last night, because the streets are wet this morning.
> *Why do you think that means it rained last night?*
> Whenever we get this far into June, it always rains at night.

The problem is obvious. But testing other arguments can be harder:

> Since 1950, the U.S. Forest Service has wasted millions trying to prevent fires. Despite the millions spent on prevention, the number of fires in western forests has remained unchanged since 1930.

This sounds reasonable, but how can we predict whether readers will also think it reasonable? We must decompose the argument and test it. There are three steps:

- **Step 1:** Infer the warrant and state it in two parts, one stating the kind of evidence it admits, the other the kind of claim it allows.

When a government agency spends money to prevent natural disasters but they occur at the same frequency$_{evidence\ side}$	the agency has wasted the money.$_{claim\ side}$

- **Step 2:** Align the evidence from the argument under the evidence side of the warrant and the claim under the claim side.

When a government agency spends money to prevent natural disasters but they occur at the same frequency.$_{evidence\ side}$	the agency has wasted the money.$_{claim\ side}$
The Forest Service has spent millions to prevent fires, but they occur at the same frequency.$_{evidence}$	The Forest Service has wasted the money.$_{claim}$

- **Step 3:** Determine whether the evidence offered is the *kind* admitted by the warrant and whether the specific claim is the kind it allows. The major terms in the evidence should match those in the warrant but be more specific.

The evidence part of the warrant refers to *general* evidence about • *government agencies,* • *spending money,* • *preventing natural disasters,* • *with no changes in frequency*	The claim part of the warrant allows claims referring to *agencies in general wasting money.*
The specific evidence refers to • *a* specific *agency* (the Forest Service), • *spending a* specific *amount of money* (millions), • *failing to prevent a* specific *disaster* (forest fires) • *with no change in frequency of fires*	The specific claim refers to a *specific agency* (the Forest Service) *wasting specific money.*

Since the evidence and claim seem to match the corresponding parts of the warrant, we can conclude that this argument establishes a valid relationship between them (though some might plausibly argue that if that warrant were left to stand unqualified, it would be false).

Here, on the other hand, is a subtly flawed argument, one about the effect of TV violence on children:

> Few doubt that when we expose children to examples of courage and generosity, we influence them for the better. How can we then deny that when they constantly see images of malev-

olent violence and sadism they are influenced for the worse? All
our data indicate that violence among children 12–16 is rising
faster than among any other age group. We can no longer ignore
the conclusion that television violence is one of the most destructive influences on our children today.

To diagnose what is wrong here, we break the warrant into its two
parts, and then align the evidence and claim under them.

When children constantly see images of malevolent violence and sadism.$_{evidence\ side}$	those children will be influenced for the worse.$_{claim\ side}$
Data show that violence among children 12–16 is rising faster than among any other other age group.$_{evidence}$	Violence on television is one of the most destructive influences on our children today.$_{claim}$

Even if each part of this argument is *true,* the argument is still
invalid, because its warrant admits neither its evidence nor its claim.
The evidence is not the *kind* of evidence the warrant allows, evidence that must refer to children "constantly see[ing] images of
malevolent violence and sadism." Nor does the specific claim match
the kind of claim allowed by the claim side of the warrant.

To fix this argument, we first have to make the evidence fit
the warrant and then restate the claim:

Few doubt that when we expose children to stories of cour-
age, compassion, and generosity, we influence them for the better. How then can we deny that when a medium such as television constantly exposes them to images of malevolent violence
and sadism, it must influence them for the worse? All our data
indicate that violence among children 12–16 is rising faster than
among any other age group. This violence results from many
factors, but we can no longer ignore the conclusion that because
television is the major source of children's images of violence, it
must be a major cause of that violence.

When a medium constantly exposes children to images of malevolent violence and sadism,_evidence side_	that medium will influence them for the worse._claim side_
TV is a child's major source of images of violence._evidence_	Television is a major cause of children's violence._claim_

The evidence and claim now seem to be the kind the warrant admits.

But a keen reader may not let the discussion end there. Even though this argument now seems formally correct, she might still object:

> Hang on. Your evidence does not, in fact, fit your warrant. Your evidence is true—images of violence do appear on television. But I don't believe that those images are "malevolent" or "sadistic." Therefore, your warrant cannot admit your evidence because your evidence is too general for the specific kind of evidence that your warrant admits. Furthermore, your claim—"major cause of violence"—is more extreme than "influence for the worse." It is too specific and so goes beyond the claim your warrant allows.

Now we see why important issues are so endlessly contestable, why when you might feel that you have constructed a watertight proof of your case, your readers may still say, Wait a minute. What about . . . ? I don't agree that your evidence counts as . . . Readers not inclined to accept your claims will question the reliability of your evidence, the truth of your warrant, and its relevance to your specific argument. Then they will debate fine points like these.

And we have not even considered those instances where there may be competing warrants, both entirely legitimate in isolation:

> When we want to express ourselves in public, we have a right to do so.
> When we are in public, we have a right not to be bothered by someone who behaves in a way that intrudes on our privacy and personal space.

Which of these warrants applies to panhandlers? to street-corner orators using loudspeakers? to street musicians? to the mentally disturbed? to people yelling at others as an act of protest? What evidence could we offer to prove either warrant? What higher-order warrants would admit that evidence?

Whenever you construct an argument, you have to offer readers evidence that *they* will consider reliable in support of a claim that *they* will judge specific and contestable. But even when your evidence is sound, your claims significant, and your warrants true, you still have to anticipate that your readers will balk if they have a deeply held assumption that does not allow them to associate your evidence with your claim.

When you start thinking about the kind of argument that you will have to make, step back and ask yourself what kind of evidence and warrant will it take to convince your particular readers. It is not enough that you think you have an airtight, open-and-shut, 100% solid case. Start with your beliefs, but remember that you have to end with those of your readers: What kind of argument will *they* accept? What kind will they reject? Let the answers to those questions shape *your* argument.

Life is too short to test every argument, but test your most important ones from your readers' point of view. Unfortunately, as usual with such advice, the trick is knowing which arguments to test. It's like knowing which words to look up in a dictionary. The words that trip you up are the ones you think you know how to spell, but in fact don't. In the same way, arguments that seem most obvious often need testing most carefully.

The more your argument asks readers to change their minds, the more significant it will seem to them (and the more convincing it will have to be). So the strongest arguments you can make are those that challenge not just the claims and evidence that your research community accepts, but the warrants that underlie them. No argumentative task is harder, because you must ask readers to change not just *what* they believe, but *why* and *how* they believe it.

If you construct an argument that challenges your readers' warrants, you have to understand what stands behind them. Recall that most warrants are claims from "higher-order" arguments. In that role, they have their own supporting evidence (along with their own, still higher-order warrant). If you know what kind of evidence backs up a warrant, you can know how best to challenge it. But the backing for some warrants is not a simple argument but a larger and more complex set of beliefs.

Before you can challenge a warrant, you first have to unpack it to understand what backs it. For example, an economist might argue,

> The population of Zackland must be controlled$_{claim}$ because it is growing too large for its resources.$_{evidence}$

Asked for his warrant, he might say:

> When a population grows beyond its resources and cannot support itself, only a reduction in population will save the country from collapse.

Were he challenged as to the truth of his warrant, he might offer as evidence particular examples:

> When the population of countries A, B, C, . . . exceeded their means, each collapsed. Therefore, we can conclude that, in general, when societies reach a point where size exceeds resources, they collapse.

Someone might reply that Zackland's population should not be reduced because it would be wrong to do so. Challenged, she might offer a warrant like this:

> Whenever a person or group discourages married couples from having children, that person or group does an inherently evil thing.

Asked for evidence to back this warrant, she would point not to quantitative data but to a system of religious or moral principles.

A third person might agree that population control is a mistake, but offer a different warrant:

> Whenever we put our minds to a problem of limited resources, we can solve it.

This warrant has a different kind of backing yet, one derived from a general pattern of cultural thinking that we are all assumed to know or believe.

These are three different and competing warrants. Each is backed by evidence that is different in kind: numbers of examples, a system of revealed truth, or inherited belief. In order to challenge any of these warrants, you have to challenge its specific kind of backing. (In the same way, be alert when you read to the different kinds of warrants on which your sources rely.)

KINDS OF WARRANTS AND KINDS OF CHALLENGES

Here is a list of the most common kinds of warrants and the kinds of backing you have to address to challenge them. They are listed in order from the easiest to challenge to the most difficult.

1. Warrants Based on Empirical Experience

These are warrants that we infer from accumulated experience. Asked to defend them, we would refer to firsthand experience, to reliable reports by others, or to the accumulated wisdom of the ages. Some are based on systematic research that produces explicit evidence:

> When certain insecticides get into the ecosystem, shells of bird eggs become so weak that fewer chicks hatch, and the bird population falls.

Some are based on inexplicit expertise developed over time:

> When a person appears in my office with X symptoms, that person is likely to have Y condition.

Some are drawn from the experience of everyday life:

> Where there's smoke, there's fire.

Challenges: Since these warrants are backed by large amounts of evidence, much of it based on experience, you must challenge its quality. You have to present contrary evidence to show that the warrant is false, or at least not entirely reliable. Since these claims are already accepted by your readers, you have to find data that are better than the accepted backing of the warrant.

2. Warrants Based on Authority

We believe some people simply because of who they are. When we respect someone by virtue of expertise, position, or person, we accept what that person says even when it contradicts the evidence of our own experience.

> When X says Y, Y must be so.

Challenges: To challenge this kind of warrant, you must challenge authority, always a risky move. Generally, you have to present two connected arguments: first, you must provide evidence that Y is not so, and second, that on this matter at least, the authority should not be believed—because the matter is beyond the reach of the authority's expertise or because the authority was not aware of your evidence. Sometimes the challenge must go even deeper: the "authority" should never have been thought authoritative in the first place.

3. Warrants Drawn from Prior Systems of Knowledge and Belief

We borrow these warrants from preexisting *systems* of definitions, principles, or theories. They are strongly held because they carry the accumulated authority of the coherence of their system. Some examples:

> *From mathematics:* When we add two odd numbers, we get an even one.

From law: When we drive without a license, we commit a misdemeanor.

From religion: When we use God's name in vain, we commit a sin.

From standard definitions of words: When a creature has feathers and wings, it is a bird.

Challenges: When you challenge these warrants, "facts" prove largely irrelevant. You must either challenge the integrity of the system, always a difficult task, or show that the instance does not fall under the warrant: *what about driving in my driveway? what counts as "in vain"? what counts as feathers? as wings?*

4. General Cultural Warrants

These are warrants that we inherit from the "common knowledge" of our culture. Some are backed by empirical experience, but most are not:

When people eat too much chocolate, they get pimples.
Early to bed, early to rise, makes you healthy, wealthy, and wise.
Whenever a king wants to abuse his subjects, he may.

Challenges: These warrants change over time, but slowly. Except for extraordinary, revolutionary moments, it is almost impossible to challenge them, because to do so challenges the basis of our culture.

5. Methodological Warrants

You can think of these as "meta-warrants." They are general patterns of thought that have no specific content until they are applied to specific cases. We use them to guide our reasoning when we derive substantive warrants like those above. The most important:

Generalization: When many instances of X exist under condition Y, then X will generally exist under condition Y.

Analogy: When X is like Y in some respects, then X will be like Y in other respects.

Cause-effect: When Y occurs if and only if X occurs first, then X may cause Y.

Sign: When X and Y are usually present at the same time, X is a sign of Y and Y is a sign of X.

Categorization: When X is a kind of Y, X will have the features of a Y.

Challenges: Philosophers and logicians have questioned even these warrants, but in matters of practical argumentation we challenge only their application or point out limiting conditions—*Yes, we can analogize X to Y,* but not if . . . (see Chapter 10).

6. Articles of Faith

Finally, there is a kind of warrant beyond warrants: Jefferson invoked it when he wrote, "We hold these truths to be self-evident . . ." This warrant is backed by the direct experience of truth:

> Whenever a claim is directly experienced as revealed truth, that claim is true.

This is the kind of truth that for some brooks no denying. It is a statement of faith, requiring no evidence.

Chapter Ten

Qualifications

This chapter discusses matters that are not difficult and can help all re-searchers, beginning and advanced, convince their readers that they are as thoughtful and judicious as they should be.

10.1 A Review

Before we turn to the art of qualifying claims, we should review the three elements necessary for every argument.

10.1.1 Claims and Evidence

To create an argument, you must state two elements explicitly:

- You must make a claim that is substantive and con-testable.
- To support that claim, you must offer evidence that is both reliable and relevant.

Evidence and claim can appear in either order:

> Toward the end of his second term, President Franklin D. Roose-velt was regularly attacked by newspapers for promoting social-ism.*evidence* Though he is today revered as one of our most admired historical figures,*context* he was at that time apparently not popular among the middle class.*claim*
>
> Today, Franklin D. Roosevelt is revered as one of our most admired historical figures,*context* but toward the end of his second term, he was apparently not popular among the middle class.*claim* He was regularly attacked by newspapers, for example, because they believed he was promoting socialism.*evidence*

In most arguments, your evidence will be new to your readers, so you have to explain it to them, breaking it into subordinate claims, which you must support with more evidence—evidence supporting evidence. In the example about Roosevelt, the evidence of his unpopularity is newspaper attacks for promoting socialism. But readers are likely to see that evidence as another claim and ask the perfectly reasonable question, *What is the evidence for your claim that newspapers attacked Roosevelt specifically for promoting socialism?*

> Today, Franklin D. Roosevelt is revered as one of our most admired historical figures,*context* but toward the end of his second term, he was not popular among the middle class.*claim* He was regularly attacked by newspapers, for example, because they believed he was promoting socialism.*evidence/claim* **In 1938, 70% of Midwest newspapers accused him of wanting the government to manage the banking system . . .** *further evidence*

You have to support your claims with evidence; but you often have to treat your evidence as subclaims that themselves also require support.

10.1.2 Warrants

The third element, your *warrant,* lets you connect a particular claim to particular evidence *validly.*

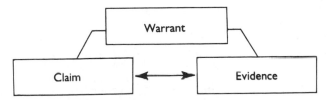

As we said in Chapter 9, when you write as an insider to other insiders, you rarely state all your warrants, but you help both your readers and *yourself* if before you draft you test your major ones. In our example, the warrant would seem to be a general belief about the role of newspapers in reflecting public opinion:

> When newspapers attack an American public official for promoting socialism, that official is in trouble with middle-class voters.

We rarely express warrants so explicitly and formulaically, preferring to imply them:

> Today, Franklin D. Roosevelt is revered as one of our most admired historical figures,*context* but toward the end of his second

term, he was not popular among the middle class.$_{claim}$ He was regularly attacked by newspapers, for example, because they believed he was promoting socialism,$_{evidence/claim}$ **a sign that a modern administration is in trouble with literate voters.**$_{warrant}$ In 1938, 70% of Midwest newspapers accused him of . . . $_{further\ evidence}$

If you build your arguments out of these three elements, you give your readers good reason to change their minds.

10.2 QUALIFYING YOUR ARGUMENT

If, however, you build your arguments out of these three elements *alone,* you may have a problem, because many readers will think such a bare-bones argument is simple to the point of simple-mindedness. Beginning researchers tend to offer arguments in a flat-footed, unqualified way, both because they think that the best argument is least in need of qualification, and because they do not recognize their own limitations. And so they write,

> Franklin D. Roosevelt was unpopular during his second term for three reasons: First, . . . Second, . . . Third, . . . Therefore, as we can see, Roosevelt was unpopular . . .

It is the standard five-paragraph argument—blunt, unsophisticated, without nuance. It succeeds only with equally unsophisticated readers.

Every contestable claim encourages readers to question the conditions in which the claim holds true and the limits of its certainty. What's more, a significant claim almost always depends on assumptions that are true only in certain circumstance. Rarely can you propose an argument whose truth is 100% certain 100% of the time.

Moreover, few readers want to read arguments that charge blindly ahead toward an unqualified conclusion, the kind that says, *Get out of my way or get run over.* They expect you to acknowledge your legitimate uncertainty, the limits of your warrant, and their legitimate questions and reservations. When you do so, you show them that you are aware of their concerns and respect their critical powers. Though it may seem paradoxical, your argument gains rhetorical strength when you acknowledge its limits.

To that end, in this chapter we add to our model of argument a fourth component, those elements that acknowledge objections and the limits of your certainty.

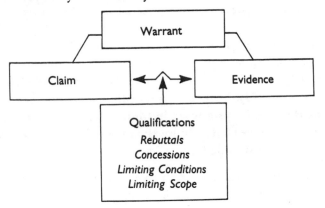

We will discuss four ways you may have to qualify your argument:

1. Rebut mistaken objections to your evidence or warrants.

2. Concede objections that you cannot rebut.

3. Stipulate conditions that qualify your evidence or limit the application of your warrant.

4. Stipulate the degree of certainty of your evidence, warrant, or claim.

10.2.1 Address Objections

Though we might all wish for readers who finish reading our report with an enthusiastic *Yes!* we know better. Reading is not like filling an empty jug with information. Engaged reading has the give-and-take of lively conversation, with readers nodding at some points, shaking their heads at others: *Wait a minute! What about . . . ?* —the kind of reading you should be doing with your sources. When you construct your own argument, you must acknowledge your readers by anticipating such questions and making the limits of your claims explicit.

Readers are most likely to question the quality of evidence or warrants. How you rebut those objections depends on their nature. For example, if you suspect that a reader might consider your evidence insufficient because she knows about evidence that runs against your claim, you must show that you have considered that additional evidence, but rejected it for good reason:

Today, Franklin D. Roosevelt is revered as one of our most admired historical figures, but toward the end of his second term, he was not popular among the middle class. Newspapers, for example, attacked him for promoting socialism, a sign that a modern administration is in trouble with middle-class voters. In 1938, 70% of Midwest newspapers accused him of wanting the government to manage the banking system. . . . **Some have argued otherwise, including Nicholson (1983, 1992) and Wiggins (1973), both of whom offer anecdotal reports that Roosevelt was always in high regard, but these reports are supported only by the memories of those who had an interest in deifying FDR.**

Or, if you anticipate an objection that your warrant is flawed, you can show why you believe that it's sound:

Today, Franklin D. Roosevelt is revered as one of our most admired historical figures, but toward the end of his second term, he was not popular among the middle class. Newspapers, for example, attacked him for promoting socialism. In 1938, 70% of Midwest newspapers accused him of wanting the government to manage the banking system. . . . **Even though Tanaka (1988) has shown that newspapers often set out to create rather than reflect public opinion, attacks as widespread as these are a reliable sign that a modern administration is in trouble with middle-class voters. Several studies have shown reliable correlations between editorial position and popular opinion . . .**

Shrewd researchers welcome such objections, even seek them out not only to improve their chances of being right but to signal readers that they are familiar with others who have worked on the same problem and come to different conclusions. When you entertain objections, you avoid overstating claims and are more likely to check whether you have enough evidence while you are still assembling your sources, not the night before the report is due.

There are four kinds of objections that you should deliberately seek out, three of which you must specifically address, one that you can choose to discuss or ignore.

1. Consider raising objections and alternatives to your claims

that, during the course of your research, *you* considered, but rejected.

You don't have to raise these, because readers are unlikely to be aware of them, but sharing them is a way to invite readers into the conversation. You shouldn't recount every blind alley and false trail. Rather, highlight the strengths of your case by raising and rebutting plausible but mistaken claims. You can seem especially judicious if you reject evidence that seems to support your claim but that you know is unreliable. By rejecting evidence that others less careful might accept, you enhance your own credibility.

2. Anticipate objections that you know readers will bring with them.

You must anticipate objections that are based on a well-known argument that contradicts some aspects of yours or that arises because you need a warrant that you know your readers won't accept. If you fail to address their objections before they think of them, you will seem ignorant of work in your field or contemptuous of their beliefs.

3. Anticipate alternatives that your readers might think of.

Your audience may not specifically reject an explanation that you offer, but they might think of alternative explanations that they believe you should have at least considered. Think of alternatives, explain them, and, if you can, refute them.

4. Anticipate objections that might occur to your readers as they read.

These objections are the most difficult to anticipate, but the most important: At some point, evidence that seems solid to you may seem dubious to your readers, or you may take a step that stretches your logic. In such cases, if you fail to anticipate objections, you will seem ignorant of the limits of your own argument and indifferent to the critical judgments of your readers. Other than disagreeing with matters of fact—with the accuracy or precision of your evidence—readers are most likely to object on these four grounds:

• You have defined key terms incorrectly.

You must be certain that your readers will agree with your definitions, because your definitions are among your systematic warrants (see p. 129). If you are researching addiction, for example, ask yourself, *When executives from cigarette companies say* smoking is not addictive, *are they denying a fact or do they define addiction in a way different from those who claim otherwise?* Well before you start to draft your argument, determine whether your readers will understand your central terms as you do. Remember that definitions are always in the service of a goal. Stipulate definitions that promote your claim.

• You have oversimplified causes and effects.

Few effects have a single cause and few causes have a single effect. If you argue that X causes Y, you can be certain that someone will object, *Wait a minute, X causes Y, but only if C, D, and E also occur, but not if Z is present, and moreover A and B also cause Y under the right circumstances.* Avoid simple answers to complex questions.

• You have overgeneralized on too little evidence.

We addressed this matter when we discussed the *sufficiency* of your evidence (p. 100). You will almost inevitably overgeneralize, simply because there are not enough hours in the day to collect all the data that you need to make a reliable generalization. All you can do is gather what you can and report it. In fact, experienced researchers seldom expect to prove anything with 100% certainty, because they cannot possibly find every bit of evidence available in the world. They can only offer their claim and invite their readers to offer evidence that disconfirms it.

• You haven't considered counterexamples and special cases.

Since readers will always try to think of counterexamples to any generalization, you should try to think of them first. If the ones you think of are aberrant or marginal cases, you can simply acknowledge that there are indeed counterexamples, but assert that they do not seriously qualify your generalization.

The easiest way to discover objections like these is with the help of a teacher, friend, or colleague. Ask anyone to play the role of contentious reader, disagreeing with what seems even slightly dubious. Ultimately, though, the responsibility is yours. If you were paid to refute your own case, what could you say? Say it and then rebut it.

10.2.2 Concede What You Cannot Rebut

Some objections you may not be able to answer. But if you are building an honest argument, you must acknowledge them. When you do, you risk revealing a possibly fatal gap in your reasoning, but you gain the advantage that comes from candidly acknowledging

your limits. You must, of course, believe that the balance of your
support more than compensates for the objection.

> Today, Franklin D. Roosevelt is revered as one of our most
> admired historical figures, but toward the end of his second
> term, he was not popular. Newspapers, for example, attacked
> him for promoting socialism, a sign that a modern administration
> is in trouble with middle-class voters. In 1938, 70% of Midwest
> newspapers accused him of wanting the government to manage
> the banking system. . . . Some have argued otherwise, including
> Nicholson (1983, 1992) and Wiggins (1973), both of whom offer
> anecdotal reports that Roosevelt was always in high regard, but
> these reports are supported only by the memories of those who
> had an interest in deifying FDR. The widespread attacks in news-
> papers across the country demonstrate significant dissatisfaction
> with his presidency. **Admittedly, the same newspapers
> praised his efforts to overcome unemployment.** But the
> evidence indicates that were it not for World War II, Roosevelt
> might not have been elected to a third term.

If you discover unrebuttable objections early, you can revise
your argument, maybe even your claim. If it happens late, you
have a problem. You could ignore the objection, hoping that your
readers won't notice. But if they do, you have a bigger problem
because they may think that you were unaware of the objection or,
worse, that you tried to hide it. If you have no good answer,
candidly acknowledge the objection as a "problem" that needs more
study, or show that the preponderance of other evidence renders
it minor.

Experienced researchers and teachers understand that truth is
always complicated, usually ambiguous, always likely to be contested. They will think better of your argument and of you if you
acknowledge its limits, especially those limits that squeeze you more
than you might wish. Concession is another way to invite readers
into the conversation.

10.2.3 Stipulate Limiting Conditions

There is another kind of objection that researchers cannot rebut
and usually do not bother to. It is a reservation about unpredictable
changes in conditions that you assume will not occur, but might.

We will win more games this year, *provided we don't suffer injuries.*

We can conclude the earthquake occurred in central Costa Rica, *so long as the instrumentation had been accurately calibrated.*

Writers are often silent about limiting conditions, especially those which state that people and things should behave as we expect. You will often hear sports commentators add a proviso about injuries to their predictions, because injuries are a common and expected part of many sports. But only rarely will scientists stipulate that their claims depend on their instruments working properly, not only because that is so obvious but also because everyone expects them to ensure that their instruments will work right.

Occasionally we stipulate a reservation, either to signal caution or to guard against foreseeable and plausible possibility:

> Today, Franklin D. Roosevelt is revered as one of our most admired historical figures, but toward the end of his second term, he was not popular. Newspapers, for example, attacked him for promoting socialism, a sign that a modern administration is in trouble with literate voters. In 1938, 70% of Midwest newspapers accused him of wanting the government to manage the banking system. . . . Some have argued otherwise, including Nicholson (1983, 1992) and Wiggins (1973), both of whom offer anecdotal reports that Roosevelt was always in high regard, but these reports are supported only by the memories of those who had an interest in deifying FDR. **Unless it can be shown that the newspapers critical of Roosevelt were controlled by special interests,** their attacks demonstrate significant dissatisfaction with Roosevelt's presidency. Admittedly, the same newspapers praised his efforts to overcome unemployment. But the evidence indicates that Roosevelt would not have been elected to a third term, were it not for World War II.

10.2.4 Limit the Scope and Certainty of Your Claim and Evidence

Even after you have rebutted every significant objection, you can seldom in good conscience assert that you are 100% certain that your evidence is 100% reliable and your claims are unqualifiedly true. Your credibility requires that you limit the scope of your arguments by hedging the certainty of your claims and evidence with modifying words and phrases.

Today, Franklin D. Roosevelt is **widely** revered as **one of** our most admired historical figures, but **toward** the end of his second term, he was not **especially** popular **among those most likely to vote.** Newspapers, for example, **often** attacked him for promoting socialism, a **good** sign that a modern administration is in trouble with middle-class voters. In 1938, 70% of Midwest newspapers accused him of wanting the government to manage the banking system. . . . Some have argued otherwise, including Nicholson (1983, 1992) and Wiggins (1973), both of whom offer anecdotal reports that Roosevelt was always in high regard, but these reports **tend to be** supported only by the memories of those who **may have** had an interest in deifying him. Unless it can be shown that the newspapers critical of Roosevelt were controlled by special interests, their attacks demonstrate **significant** dissatisfaction with **key aspects of** his presidency. Admittedly, **many of** the same newspapers praised his efforts to overcome unemployment. But the **weight of** evidence **suggests** that were it not for World War II Roosevelt **probably** would not have been elected to a third term.

The words and phrases that limit your evidence and claims give your argument nuance.

You need not state each moment of uncertainty, only the important ones. If you hedge too much, you will seem timid or uncertain. But in most fields, it is foolish to avoid every "seem," "may," and "probably" in some vain hope that readers will be impressed by flat-footed certainty. Some teachers strike out all hedges. *Don't say that you* believe *or* think *something is so. Just say it!* But what most of them don't like are hedges that qualify every trivial claim. And it must be acknowledged that, in some fields, hedging is considered more objectionable than in others. Teachers and editors who condemn all hedging are just wrong about the way most careful researchers report their results. Every researcher must know how to seem thoughtfully confident, which means knowing how to express the limits of that confidence.

All of these points implicitly address what we call your *persona* or *ethos*—the picture of your character that readers infer from your style of writing and thinking. Few elements more significantly influence how they judge your character than how you handle uncertainties and limitations. It takes a deft touch. Hedge too much and you seem mealymouthed; too little, smug. Unfortunately, the line between hedging and fudging is thin. As usual, watch how those in your field manage uncertainty, then do likewise.

10.3 BUILDING A COMPLETE ARGUMENT

Here again is the whole structure:

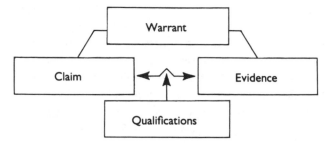

Remember that the arrows indicate only logical relations, not any *necessary* sequence in any particular argument in real time. Real-time arguments are almost always less sharply defined, more discursive, less linear. Warrants get folded into the same sentence as a claim; a reservation creeps in as a parenthetical aside; several lines of argument converge on a single claim. Most important, a large and complex argument is built out of simple arguments of different kinds that depend not only on different warrants but different kinds of warrants. Regardless of these surface differences, all responsible arguments are built from these four elements.

You can begin any individual unit of argument with a claim or conclude with it; you can rebut objections at the beginning of an argument, in the middle, just before the final claim, or even after. Suppose we now arrange the elements of the "same" argument in two different ways. In the first example, the argument begins with a straightforward statement of the claim (boldfaced) and evidence (underlined), then qualifies it (italicized) and rebuts objections (capitalized). The second lays out the qualifications and rebuttals first, and then works toward the claim. As you can see, the rhetorical effects are quite different:

> Although today Franklin D. Roosevelt is revered as one of America's more admired historical figures, **there is evidence to suggest that, at the time, he was not as popular as many**

now claim; indeed, that had it not been for WWII, he might not even have been elected to a third term. In the late 1930s, many newspapers attacked him for promoting socialism, a sign that any modern administration may be in trouble with the people, *or at least among the most literate segments.* In 1938, for example, 70% of the newspapers in the Midwest accused him of wanting the government to manage the banking system. . . . *Unless these newspapers were controlled by special interests,* **their attacks demonstrate that Roosevelt was not as widely admired as some have recently suggested.** *Admittedly, these same newspapers often praised him for his attempts to end unemployment.* BUT THOSE WHO ARGUE THAT ROOSEVELT WAS WIDELY LOVED (NICHOLSON 1982, WIGGINS 1973) CONCENTRATE TOO HEAVILY ON THE MEMORIES OF THOSE WITH AN INTEREST IN DEIFYING HIM. The most reliable evidence suggests that Roosevelt was far from being admired by all of the people.

In the late 1930s, newspapers praised Franklin D. Roosevelt for his attempts to end unemployment, and some have recently argued that in his time he was widely loved. (Nicholson 1982, Wiggins 1973). Indeed, today, Roosevelt is revered as one of America's most admired historical figures, but THOSE WHO CLAIM HE WAS *WIDELY* LOVED MAY CONCENTRATE TOO HEAVILY ON THE MEMORIES OF THOSE WITH AN INTEREST IN DEIFYING HIM. In fact, many of those same newspapers that praised him attacked him for promoting socialism, a good sign that any modern administration may be in trouble with the people, or at least among its most literate segments. In 1938, for example, 70% of the newspapers in the Midwest accused him of wanting the government to manage the banking system. . . . *Unless these newspapers were controlled by special interests,* their attacks demonstrate that **Roosevelt was not as widely admired as some have recently suggested. In fact, some evidence suggests that, had it not been for World War II, Roosevelt might not have been elected to a third term.**

10.4 ARGUMENT AS A GUIDE TO RESEARCH AND READING

The structure of an argument is invaluable in helping you think through your project, from beginning to end.

1. Its elements can guide your research. If you can anticipate what you have to include in your report—not just claims and evidence but warrants and qualifications—you can read accordingly, looking not only for support, but for disagreements to rebut.

2. The elements of an argument help you read more critically. As you read your sources, you should ask the same questions that your readers will probably ask you:

Your Questions	Your Source's Answers
What is your point?	*I claim that . . .*
What is the scope of your claim?	*I limit it to . . .*
What evidence do you have?	*I offer as evidence . . .*
What links evidence to claim?	*I offer this principle . . .*
But what about . . . ?	*I can rebut that. First, . . .*
But what if . . . ?	*My claim stands so long as . . .*
*No problems here **at all**?*	*Well, I have to admit that . . .*

3. These elements can help you organize your materials and your thinking as you prepare for your first draft. Your earliest outlines should focus on the elements of your argument.

4. The elements of your argument can help you identify the parts of your paper and guide your drafting.

5. And finally, the elements of your argument can help you anticipate what your readers will think of *you,* because nothing more reveals a person's character than the way that person tries to convince others to change their minds.

10.5 A Few Words about Strong Feelings

In the last few chapters, we have laid out a model of argument that emphasizes cool logic. In many fields—the natural sciences, for example—nothing is valued more highly than an argument that moves from reliable evidence to a significant claim in ways that are patient, dispassionate, and above all logically sound. But all readers respond to—and with—more than cool reason when in a sound argument they also sense a researcher's warm commitment to what he believes is the truth. When readers hear in an argument not only the voice of reason but accents of its commitment, or even passion when passion is called for, those readers attend to that argument more closely than they might to another that seems as intellectually sound, but bloodless. It is a matter that no discussion of argument can ignore.

But it is a quality of discourse almost impossible to teach directly. When you evaluate the logic of an argument, its *logos,* you can look for its parts out there on the page, reconstruct those that you can't find, study their relationships, and then decide whether the writer has earned your assent. As you analyze someone else's reasoning in this way, you engage in the same kind of reasoning that you are analyzing and evaluating, and if your analysis is challenged, you can analyze your own reasoning in the same way you did the writer's. On the other hand, when you evaluate the strength of that writer's personal commitment to her claim, you have as hard evidence only a response that is immediate, unselfconscious, and visceral. Only from your own felt response to the *pathos* of an argument, a response that to others can seem non-logical, or even illogical, can you decide whether to *trust* that writer as, in fact, sincere. The line between faked and genuine sincerity is hard to discover. As the cynical joke goes, when you can fake sincerity, you have it made.

If we knew a sure way either to detect insincerity or to ensure that your readers will hear in your prose a sense of your own true commitment, we would tell you. But we can't. We can only echo what rhetoricians since Aristotle have said: Every argument depends on three appeals—its *logos* (logic), its *pathos* (emotional quality), and its *ethos* (the writer's perceived character), and our readers' conviction is woven of all three.

QUICK TIP:
ARGUMENTS—TWO COMMON PITFALLS

Arguments fail for many reasons, but for inexperienced researchers the two most common are the following.

INAPPROPRIATE EVIDENCE

If you are working in a new field and are not yet familiar with its characteristic modes of argument, it is easy to fall back on the forms of argument you already know. Every time you enter a new field, you must find out what's new and different about the *kinds* of argument your teacher expects you to make. If you learned in your first-year composition class to search for evidence in your own experience and then on the basis of those memories offer opinions on issues of social concern, do not assume that you can rely on the same process to create persuasive arguments in fields that emphasize "objective data," such as experimental psychology. On the other hand, if as a psychology or biology major you learned to gather data, subject it to statistical analysis, and avoid attributing to it your own feelings, do not assume that you can use the same method to construct a good argument in art history.

This does not mean that what you learn in one class is useless in another, only that you must watch for differences among fields. You have to be flexible enough to adapt to what's new about a field, and at the same time trust the skills you already command. You can anticipate this problem as you read by noting the kinds of evidence writers offer to support their claims. Here are some different kinds of evidence to watch for in different fields:

- Personal beliefs and episodes from writers' own lives, as in a first-year composition course.
- Detailed documentary data assembled into a coherent story, as in some kinds of history.
- Fine-grained descriptions of everyday behavior, as in anthropology.
- Quantitative summaries of social groups, as in sociology.

- Quantitative data aiming at a single result, as in engineering.
- Direct quotations, as in most of the humanities.
- Networks of shared meanings imposed on a seemingly disparate body of citations, as in literary criticism.
- Networks of principles, implications, inferences, and conclusions independent of factual data, as in philosophy.
- Citations and borrowings from previous writers, as in the law.

Just as important, note what kinds of evidence *never* appear in the arguments of your field. Anecdotes enliven sociological explanations but usually do not count as good evidence; fine-grained narratives of events in a laboratory do not count in physics; networks of logical principles and conclusions do not do the whole job in chemical engineering.

COMFORTABLE SIMPLICITY

When you are new to a field, everything can seem confusing. Like everyone else in those circumstances, you will look for simplicity—a familiar method or an unambiguous answer, any simplification that can help you manage complexity. And once you find it, you are likely to oversimplify. As you begin your research, be aware that no complex effect has a single, unambiguous cause; no serious question has a single, unqualified answer; no interesting problem has a single, simple methodology to solve it or a single solution. Seek out qualifications; formulate at least one alternative solution to your problem; ask whether someone else in the field approaches your problem differently.

Once you become familiar with the methods of inquiry in a field, its typical problems, schools of thought, and so on, you will begin to see its logical and conceptual structure. But as you learn more, you will discover a second kind of complexity, the complexity of competing solutions, competing methodologies, competing goals and objectives—all marks of a lively field of inquiry. The more you learn, the more you recognize that, while things are not as blindingly complex as you first thought, neither did they turn out to be as simple as you then hoped. It is at this point that the beginning researcher succumbs to another kind of overgeneralization. Once you learn how to construct one kind of argument in that field, you try to make that same argument over and over. Be aware that circumstances always differ, that though data from one case might look like data from another, they are likely to be different in surprising ways.

Part Four
Preparing to Draft, Drafting, and Revising

Prologue: Planning Again

NO FORMULA CAN TELL YOU when to start drafting. Booth begins "too early"; then once his ideas become clearer, he faces the nasty problem of throwing away most of what he's written. Colomb is an inveterate outliner, producing a dozen outlines and two or three "advance summaries." Williams tries out as many versions as do Booth and Colomb, but in his head; he writes as he goes, but he starts a serious draft only when he has some sense of the whole.

PREPARING FOR THE FIRST DRAFT

We can offer no trick for knowing when *you* should start drafting, but you can prepare for that difficult moment if you start writing notes and summaries and critiques from the moment you begin. You are ready to start a serious first draft when you have a plan, no matter how rough and tentative—in your head or on paper: an outline, an advance summary, or even just a general idea of its shape. That plan should reflect the following elements:

- A picture of your *readers.* What they expect; what they are likely to know or assume; what opinions they bring with them; why they should care about your problem. (Review Chapters 2 and 4.)
- A sense of the *character* you want to project. Will you present yourself as someone passionately committed to a point of view, or as a dispassionate observer who explores all alternatives before arriving at a conclusion? (Review Chapter 10.)
- A *question* that defines some gap in knowledge, some flawed understanding that you want to resolve. (Review Chapter 4.)
- Your *main claim* or *point,* and some of the *subpoints* that will support it. These may be provisional, a best guess at an answer to your question. But it is better to start

149

with a claim that you know you may later abandon
than with none at all. (Review Part III.)

• The *sequence of the parts* of your paper, the subject of
  Chapters 11 and 13. Some papers have specified parts
  in a particular order; for others, you must create your
  own structure. In either case, before you draft, decide
  on the parts you plan to have, their sequence, and how
  the information you have collected fits into them.

Before you begin drafting you should have some ideas about
these elements, but they don't have to be detailed, because you will
almost certainly develop your ideas as you write. In some papers
(for example, a lab report with a single definitive result) you might
be certain of your point and argument before you draft, but in
other papers, especially in the humanities and social sciences, you
can expect—even hope—that as you write, you will change your
point, perhaps several times, each time discovering something new
and more interesting. Writing is a way not to report what is in
that pile of notes, but to discover what you can make of it.

THE WRITING PROCESS

Just as we plan in different ways, so we have many ways to
write. But many experienced writers follow two principles.

• First, they respect the complexity of the task. They do
  not expect to march straight through the process to a
  finished draft. They know that as they work they will
  discover something new that causes them to rethink
  their project.
• Second, they know that much of their early writing
  will end up in their files or wastebasket, and so they
  start early enough to leave time for dead ends, restarts,
  new ideas, further research, and revision—especially
  revision, because they know that the really productive
  work begins after they see not what they think they
  know, but what they are finally able to say.

Then once they start to draft, they keep in mind a few more principles:

- They draft as fast as they reasonably can, leaving matters of spelling, punctuation, grammar, and so on until later.
- They collect reactions from anyone whose opinions they trust.
- Above all, long before they have reached this point, they have been writing along the way.

Even experienced writers feel that deadlines come too soon; they long for another month, a week, just a day. Some spend their whole careers on one problem, and even then they feel that they have to deliver before their ideas have matured. Writing before you feel your work is complete will always seem frustrating, but less so if you think of your paper not as a polished gem but as a stepping stone to better understanding. No researcher, not even the best, ever has the last word, fortunately for us all. If the tradition of research has taught us anything, it is that Truth has a way of changing. The best we can hope for is to make our interim report as clear, as complete, and as close to being right as we can: *After my best efforts, here is what I believe to be true—not the whole truth, but a truth important to me and to my readers, a truth that I have tried to support strongly enough and express clearly enough that they will find in my argument good reason to agree or at least to reconsider what they believe.*

QUICK TIP:
OUTLINING

An outline can be one of your most important tools, but it can also be a nuisance. The three of us remember when we were beginners being forced to produce one of those classical outlines: main headings with roman numerals, each level neatly indented, no subhead "A" without a parallel "B." (Of course we really wrote the draft first, then the outline, then claimed that we did it the other way around.)

But while formal outlines used at the wrong time are useless, most of us can start to draft only after we have an outline of some kind, no matter how sketchy. (In the next three Chapters, we'll discuss some ways to discover a useful outline.) At this point, it is enough to distinguish between a *topic-based* and a *point-based* outline and to know when each is useful.

A topic-based outline consists of a series of nouns or noun phrases:

    I. Introduction: Word Processors in the Classroom
   II. Uses of Word Processors
      a. Labs
      b. Classroom Instruction
      c. Dorm Clusters
  III. IBM vs Mac
      a. Study Methods
      b. Study Conclusion
      c. Questions about Study
  IV. Revision Studies
      a. Study A
      b. Study B
      c. Study C
   V. My Experience
  VI. Grade Survey
 VII. Conclusions

Such skeletons help in the earliest stage of your thinking and planning, but they do little to advance you from a topic to a question and on to a draft. The readier you are to write, the more you should focus your outline on your points, which will be the most important subclaims in your argument. Look above at topic IV, "Revision Studies." Does it help you predict what arguments the writer will make? A combined point-based and topic outline would do a better job:

    I. Introduction: Value of classroom word processors uncertain.

   II. Different uses have different effects.

      a. All uses increase flexibility:

         −for students (revisions, ability to try out ideas)

         −for teachers (revision assignments, paper comments)

      b. Networked computer labs allow student interaction.

      c. Classroom instruction does not enhance learning.

  III. Does IBM or Mac produce more thoughtful papers?

      a. Research methods differ.

      b. One study concludes that "graphical interface" makes students frivolous or attracts more frivolous students.

      c. Conclusions are unreliable because

         −no controls over sample

         −no distinction between "frivolous" and "creative"

         −relies too much on "image"

  IV. Studies show that the benefits on revision are limited.

      a. Study A: writers more prolix.

      b. Study B: writers need hard copy to revise effectively.

      c. Study C: spell checkers and grammar checkers lull students into a false sense of security.

   V. Studies ignore emotional stress on those who do not already use word processors.

  VI. Survey shows that better students more often use word processors.

 VII. Conclusion: Too soon to tell how much word processors improve learning.

      a. Too few reliable empirical studies.

      b. Too little history, too many programs in transition.

      c. Basic issues have not been studied.

This outline is more helpful, not just because it has more information, but also because it shows the relations among claims. With this kind of outline, you can also see better where points do or do not hang together. No less important, because each point is a *claim* in some argument, you will have to support each of them with *evidence,* and that will motivate each stage in your work. Of course, you might not be able to create this kind of outline until you have finished a draft. At that point, this kind of outline is especially useful.

Inexperienced writers often think that the only time to create an outline is just before they begin to draft. But outlines of different kinds are part of the project from beginning to end.

Points, Claims, Solutions, Answers, and Other Terms for Your Most Important Ideas

When we discussed arguments in Part III, we used the term *claim* to refer to the sentence or sentences that make the central assertion that your argument supports. We also suggested that you make an outline of your main claim and major subclaims. The order of elements in that outline of your argument may be different from the outline of your paper, but its claims and subclaims should appear in both.

When we discussed questions and problems in Part II, we also used the terms *answer* and *solution* to refer to the sentence or sentences that resolved the matter. That answer or solution will also be the main claim of your argument and the main *point* of your paper.

We have used so many terms for the same sentences because each term defines those key sentences from a different angle. Most papers, research paper or not, make *points*—first, a main point that is central to the whole paper and then subpoints that are central to each section and paragraph. The point of your paper (or a section or paragraph) is your most important idea, its bottom line, the sentence(s) that all the rest support. In a research report, your main point and your most important subpoints will also be *claims* that you support with evidence. Your main point/claim will also be your answer to your research question or the solution to your research problem.

Points have been given so many names because they are central to writing effective papers. You might also be familiar with the term *thesis.* Your main *thesis* is the same as your main *point,* which is the same as the main *claim* in your argument. Another term that you may think of is *topic sentence.* A topic sentence is usually the point sentence of just a paragraph. No serious damage is done if you think of *answer, solution, claim, point, thesis,* and *topic sentence* as meaning roughly the same thing.

CHAPTER ELEVEN

Pre-Drafting and Drafting

If you have followed our advice all along, you have already done a lot of writing that you might think of as drafting. But if you are having trouble getting started, this chapter should help, whether you are on your first or twentieth research project.

NOTHING IS EASIER THAN PUTTING OFF YOUR FIRST DRAFT—*Just another week of reading,* you think, *another day, another hour; as soon as I finish this cup of coffee, I'll be ready to think hard about getting ready to draft.* And in the long run, nothing is more certain to cause you grief. Writing is hard, certainly harder than reading more. Still, you have to start sometime, and you'll start more easily if you've been writing along the way, and if you plan carefully now.

11.1 PRELIMINARIES TO DRAFTING

We have emphasized how important it is to *plan* your project, even though you know that you may have to change it. Drafting is no different. Drafting goes faster if you have a plan than if you just sit down and try to think of the first word.

11.1.1 Know When You Are Ready

You will know you are ready to plan a serious first draft when you have a general sense of the elements we sketched in the prologue: a research question, a possible answer, and a body of evidence to support the answer. It also helps to have a scratch outline laying out a sequence of points. If you are an advanced researcher, you should also have thought about:

- the major *warrants* that your readers have to accept before they will accept your evidence and claim,
- the *objections* that you have to rebut, and those that you cannot.

A few researchers have settled ideas about every element before they draft a word, especially when their research involves quantitative analysis that produces a result requiring little interpretation:

> What is the effect of wearing a motorcycle helmet? Cyclists who wear them suffer 46% fewer serious head injuries than those who do not.

But when your paper requires you to synthesize sources, engage in conceptual analysis, interpretation, judgment, and evaluation, you may not have a clear sense of your results before you start drafting. You may not even have a clear idea of your problem. In that case, the act of drafting is what will help you analyze, interpret, judge, and evaluate.

You can expect such moments of uncertainty. How you deal with them depends on the reasons for your confusion. Your most likely problem will be that you aren't certain you have a good point to make. If so, use the previous chapters to work on your argument. Review the questions you asked. Ask them again. If you have points but are not sure that they add up to a good main point, step back for a long view to ask how all those points bear on your question. If you have three good candidates for a main point, pick one that interests you the most; or better, the one that you think will most interest your readers.

You will *know* you are ready to plan a first draft when you have enough evidence to support a point that you can articulate like this:

- It is concise enough to state in a sentence or two.
- It is contestable, not self-evident, one needing your evidence.
- It states in specific words the central substantive concepts you can develop in the body of your paper.
- It does not depend for its power on words like "interesting," "significant," or "important" and its conceptual words go beyond abstractions like "relationship between X and Y" or "the influence of X on Y."

11.1.2 Exploratory Drafting vs. Final Drafting

Before we describe the planning process, we have to say again that many experienced writers begin to draft long before they can answer any of these questions, because they are willing to invest time in a process that they believe will lead them to their answers. But they do so knowing they will cut much of what they write from their final draft. They understand that in early drafts they will only summarize sources and record speculations, false starts, and late-night thoughts. They know that their earliest draft may only slightly resemble their last one. And so they start early.

We would never discourage anyone from doing that, but the risk of exploratory writing is that you become so attached to it that you cannot give it up; worse, you may not recognize that it is only a narrative of your quest; and worst of all, a deadline may force you to make it your final draft. Exploratory drafting can help you discover things you never dreamed of, but it is not efficient if a deadline allows you only a draft or two. If you want to reach a final draft more efficiently, you have to plan more thoughtfully.

11.2 Planning Your Organization: Four Traps

Beginning researchers often have problems organizing a first draft because they are learning how to write at the same time they are discovering what to write. As a consequence, they often lose their way and grasp at any principle of organization that seems safe. There are some useful rules of thumb in planning a paper, but there are also four common principles of organization that you should *never* seize on as a first—or even second—resort.

11.2.1 Repeating the Assignment

Beginning researchers often map their papers onto the literal structure of their assignment. If your assignment lists four points to consider, organize your paper around them *only if* the assignment requires you to do so and *only if* you can think of no other way. If the assignment asks you to compare A and B, don't assume that your paper must have two halves, one for A, one for B, and in that order. And under no circumstances repeat the assignment word for word in your first paragraph, as in this example.

Instructor's Assignment:

Different theories of perception give different weight to cognitive mediation in processing sensory input. Some claim that input reaches the brain unmediated, others that receptive organs are subject to cognitive influence. Compare two or three theories of visual, aural, or tactile perception that take different positions on this matter.

Student's Opening Paragraph:

Different theorists of visual perception give different weight to the role of cognitive mediation in the processing of sensory input. In this paper I will compare two theories of visual perception, one of which (Kinahan 1979) claims that stimulus reaches the brain unmediated, the other of which (Wright 1986, 1988) claims that cognition influences the visual receptors themselves.

11.2.2 Summarizing Sources

If you are unfamiliar with a subject or a whole field, you will find it too easy to rely on your sources more than you should. You face different problems in different kinds of research.

In library research, avoid building your paper out of summaries and quotations, particularly as you draft the first half of your paper, when you provide "background." The worst expression of this impulse is called "quilting." You stitch together quotations from a dozen sources in a design that reflects little of your own original thinking. When teachers see only summaries and paraphrases, they deliver that terminal judgment, *This is all summary, no analysis.* Some fields require you to survey what others have said, but in those summaries your instructor will look for *your* angle. You cannot reserve your contribution for a few sentences at the end.

In field research, do not simply report observations or repeat quotes from interviews. Here too your contribution must appear *throughout* your paper in the principles of selection you apply to your data. If, for example, you are reporting on interpersonal interactions in a workplace, you cannot describe everything you observed. You must select and arrange your observations and interviews to reflect your *analysis* of what is relevant. Use observations to support rather than substitute for your analysis.

In laboratory research, don't bury your results in a narrative of your lab work. Your contribution must appear in a statement of your method that selects only the relevant details. Don't mix methods, results, and every misstep along the way.

11.2.3 Structuring Your Paper around Your Data

You can recognize this problem when you organize your paper around the most predictable people, places, or things in your data, rather than asking yourself whether you could reorganize your information into new categories that would more accurately reflect your claim or be more useful to your readers.

Suppose you are writing about dreams, the imagination, Freud, Jung, social variables, and biological variables. The principle of organization that some might seize on right away is first half Freud, second half Jung, because their names are the most recognizable. That order might be useful to readers who were particularly interested in Freud and Jung, but it is so predictable that it might minimize your own contribution and fail to show readers how you want them to understand the material in the context of your claim. A second principle might be first half social variables, second half biological. But if you claim that "Dreams depend more on biological variables, the imagination more on social variables," then you ought not to organize your paper around Freud and Jung or even around social and biological variables, but around dreams and the imagination.

Before you settle on your outline, spend some time categorizing and recategorizing your data as an exercise that might help you hit on the point of view most useful to your readers. What organizing categories would best reflect the categories of your claim? You might even discover a claim more interesting than the one you have tentatively proposed.

11.2.4 Structuring Your Paper around a Narrative of Your Research

Don't draft your paper as if you were narrating an archaeological dig through the records of your research. Few readers are interested in a blow-by-blow account of what you found first, then the obstacles you overcame, then the new lead you pursued, then how you hit on an answer. That kind of narrative form can creep into your paper if you keep your notes like the layers of a civilization and you write your paper by peeling them away one at a time, recording every step.

You will see signs of this kind of problem in language like *The first issue that I addressed was . . . , Then I compared . . . .* Put a question mark beside every sentence that uses language referring specifically to what you did when you conducted your research or that explicitly refers to your acts of thinking and writing. If you see more than a few such references, you are probably not solving your problem but telling a story about yourself. Cut those that do not help your readers understand your argument.

You avoid this problem by analyzing your data as you gather them.

11.3 A PLAN FOR DRAFTING

In what follows, we lay out a series of steps in an order that you should not take as fixed. Reorder them to suit your own needs, but be sure to include them all.

11.3.1 Determine Where to Locate Your Point

If you have even a vague sense of your main claim, express it and then decide where you will state it for the first time. Practically speaking, you have only two choices:

- in your introduction, specifically *as its last sentence* (not as its first), so that your readers know where you are taking them;
- in your conclusion, so that you reveal your destination to your readers only after your evidence seems inevitably to have taken them there.

This is a fundamental organizational choice, because it defines the social contract you make with your readers. If you state your main point—your main claim, the solution to your problem, the answer to your question—at the end of your introduction, you are saying to them: *Readers, you now have control of this report. You know the outlines of my problem and its solution. You can decide how—or even whether—to go on reading.*

On the other hand, if you wait until your conclusion to state your main point, you establish a very different—and more controlling—relationship: *Readers, I am going to lead you all the way through this report, considering every alternative I offer in the order I set, until the end, where I will reveal to you my conclusion.*

Most readers prefer to see the main point in the introduction of a report, specifically at the *end* of the introduction, because that organization gives them the greatest autonomy. In some fields, however, the standard forms require you to locate your main point in your conclusion. If so, remember that your readers still need to know where your paper is headed, and they will look for you to announce in the introduction some sense of direction (they may, of course, flip to your conclusion, read that, and start over or put you aside). Readers want you to start them off with a strong sense of the path ahead and then to keep them posted along the way.

The same principle applies to major sections of your paper and to their subsections. Readers begin looking for the main point of a section at the end of its introduction. If that introduction is a single sentence, then that point will be the first sentence of that section. If the introduction is longer, readers look for the main point in the last sentence of that introduction. Of course, you may have reason to put the point of a whole section at its end. But at the beginning of every section, readers still need an introductory sentence or two to lead them into its body. So even if you put your point last in a section, draft a sentence or two for the beginning that will lead your readers to the point at the end.

In general, plan your paper so that a reader following our Quick Tip on fast reading (pp. 82–84) could skim your paper and get its general gist and the gist of each section. We will return to these principles in Chapter 13, when we discuss revising organization.

11.3.2 Formulate a Working Introduction

The first thing you need in front of you as you draft is the question you are asking and a sense of its answer, something you can sketch in a few words. Starting a draft can be so hard that some of us wait until we've written the last words before we attempt any of the first ones (we devote all of Chapter 15 to your last-draft introduction). But most of us still need some kind of working introduction to point us in the right direction. We know we will discard it, but that working introduction should be as explicit as we can make it.

The least useful working introduction announces only a topic:

> This study is about birth order and success among recent immi-
> grants.

Better to open with a bit of context. Then, if you can, succinctly
state your question as a problem, followed by its solution, if you
know it. If you don't, try to characterize the kind of solution that
you might find:

> First-born middle-class native Caucasian males are said to
> earn more, stay employed longer, and report more job satis-
> faction.·*context*
>
> But no studies have looked at recent immigrants to find out
> whether they repeat that pattern. If it doesn't hold, we have to
> understand whether another does, why it is different, and what
> its effects are, because only then can we understand patterns of
> success and failure in ethnic communities.·*the research problem*
>
> The predicted connection between success and birth order
> seems to cut across ethnic groups, particularly those from South-
> east Asia. But there are complications in regard to different
> ethnic groups, how long a family has been here, and their eco-
> nomic level before they came.·*a sense of the outcome*

This introduction only sketches the problem and nods toward its
solution, but it is enough to start you on track. In your last draft,
you will revise it so that it articulates the clearest idea of the
particular problem and solution that you finally discover.

If you are really stuck for a way to begin, go back to the
beginning of Chapter 4 and use that framework:

> *I am studying* economic success and birth order among recent
> male immigrants from Viet Nam,
> > *because I want to find out whether* the same pattern that holds
> > in native-born males holds with them,
> > > *in order to understand how* different cultural forces,
> > > family influences, and so on influence their social
> > > mobility.

11.3.3 Determine Necessary Background, Definitions, Conditions

Once you have a working introduction, decide what your readers must *immediately* know, understand, or believe before they can

understand anything else. Depending on their field, many writers at this point spell out their problem in more detail than they were able to in their introduction. They define terms, review prior research, establish important warrants, set limits on their project, locate their particular problem in a larger historical or social context, and so on.

Your greatest risk here is that you will run on for several pages summarizing your sources in a way that your readers will think is beside the point. Present just enough background information from your notes so that readers not intimately familiar with your topic will understand any special terms, any research that motivated yours, and the basic facts about the material you studied. *When you begin to draft, however, you cannot let this summary dominate your paper.* It should provide only enough background information to let readers understand what follows. If your background section is more than a couple of pages long, end it with a concise summary of what you want your readers to carry with them as they begin the main body of your argument.

11.3.4 Rework Your Outline

When you begin to plan the body of your argument, don't forget that you can always arrange the elements of an argument in more ways than one (see pp. 142–143). To discover a good way, you'll need to manipulate the structures we discussed in Chapters 7–10, trying several orders. It costs less to discard bad choices now than after you begin drafting. In all these considerations, though, put your readers *first.* Try arranging your information in orders that reflect *their* needs.

In that regard, there are a few reliable principles, and all of them turn on what your readers already know and understand.

Old to New. In general, readers prefer to move from what they know to what they don't. So a good principle for ordering the body of your paper is to begin by reviewing *briefly* what your readers know so that they can move to what they will think is new. Take this principle as a general guide when you are stuck: what in your data and your argument will be most familiar to your readers, what least? Start with the familiar, move to the unfamiliar.

Shorter and Simpler to Longer and More Complex. In general, readers prefer to encounter shorter, less complex material before longer, more complex. Which elements of your argument will your readers understand most easily; which less easily?

Uncontested to More Contested. In general, readers move more readily from less contested to more contested issues. Which elements of your argument will your readers accept most easily, which might they resist

> **Finding the Right Order**
>
> Almost to the last draft of this chapter, we had located what is now Section 11.2, "Four Traps," *after* the section you are now reading. But we realized that you would understand what you *should* do more easily if we first warned you to avoid some of the mistakes students typically fall into.

most strongly? If your main claim is controversial and you can present several arguments to support it, try starting with the one your reader is most likely to accept.

Unfortunately, these criteria often pull against one another: what some readers understand best are the objections they hold most strongly; what you think is your most decisive argument can be the newest and most contested claim. We can offer no certain rules here, only variables to consider. Try these, as well:

- chronological order;
- logical order, from evidence to claim, and vice versa;
- concessions and conditions first, then an objection you can rebut, then your own affirmative evidence, and vice versa.

In short, give yourself a chance to discover the potential in what you know by testing your points in different combinations. Presiding over all your judgments must be this prime principle: What must your readers know now before they can understand what comes later?

11.3.5 Select and Shape Your Material

At this point you can expect to set aside much of your material, because it will seem irrelevant. That does not mean you wasted time collecting it. Research is like gold mining: dig up a lot of raw material, pick out a little, discard the rest. Even if all that material never appears in your report, it is the tacit foundation of knowledge on which your argument rests. Ernest Hemingway once said that you know you're writing well when you discard stuff you know is good. You know you have constructed a convincing argument when you find yourself discarding material that looks good—but not as good as what you keep.

11.4 Creating a Revisable Draft

If you think you are ready to begin putting words on paper, reflect for a moment about the kind of drafter you are (or might want to be).

11.4.1 Two Styles of Drafting

Quick and Dirty: Many writers find it most efficient to draft as fast as they can make pen or keys move. Not worrying about style or correctness, or even clarity (least of all spelling), they try to keep up the flow of ideas. If a section bogs down, they note why they got stuck, refer to their outline for their next move, and push on. If they are on a roll, they do not type out quotes or footnotes: they insert just enough to indicate what to do later.

Then if they do freeze up completely, they have things to do: fiddle with wording, add quotes, play with the introduction, review what they've drafted, in a sentence or two summarize the ground they have covered, make sure that the bibliography includes every source cited in the text. As a last resort, they correct spelling, punctuation—anything that diverts their minds from what is blocking them but keeps them on task, giving their subconscious a chance to work on the problem.

Or they go for a walk.

Slow and Clean: There are others who cannot work with such "sloppy" methods but only "word-by-perfect-word," "sentence-by-polished-sentence." They cannot start a new sentence until the one they are working on is dead right. If you cannot imagine a quicker but rougher style of drafting, don't fight it. But remember: the more you nail down each small piece, the fewer alternatives you have thereafter. You will experience your greatest difficulty if you suddenly see things in a new way and you try to make large-scale revisions. If your "sentence-by-sentence" drafting has built in careful transitions and connections among paragraphs and sections, your paper will seem like a wall of interlocking granite blocks. Even a minor change will require more collateral changes than you may want to make. For this reason, if you are a "sentence-by-sentence" drafter, you must have a detailed outline that tells you where you are going and how you will get there.

11.4.2 Create a Routine

Whichever style is yours, establish a ritual for writing and follow it. Ritualistically straighten up your desk, sit down, sharpen your pencils or boot up your computer, get the light just right, knowing that you will sit there for an absolutely irreducible period of time. If you sit staring, not an idea in your head, write a summary: *So far, I have these points* . . . . Or look at the last few paragraphs you wrote, and treat some important bit of evidence as a claim in a subordinate argument. Identify the key words in each subordinate claim, asking what evidence would encourage your readers to accept them, and start writing:

1. Many newspapers attacked Roosevelt. *What evidence shows that **many** newspapers **attacked** Roosevelt?*
2. They attacked him for promoting socialism. *What evidence shows that they attacked him for **promoting** socialism?*
3. If they attacked him, he must have been unpopular. *What evidence shows that if newspapers attack a president, he must be **unpopular**?*

Do this for each major element in your argument. Then depending on your deadline analyze each sub-sub argument in the same way.

11.5 THE PITFALL TO AVOID AT ALL COSTS

It will be as you draft that you risk the worst thing that can happen to a researcher: In the heat of drafting, you are confidently plowing through your notes, finding good things to say, filling up the page or screen with lots of good words. *And those words belong to someone else.*

Plagiarism is a topic that embarrasses everyone, except, perhaps, the successful plagiarist. Every researcher needs to give it serious thought. Some acts of plagiarism are deliberate. No one needs help to know that it is wrong to buy a term paper, copy a report from a fraternity's files, or use large chunks of an article as though the words were your own. But most plagiarism is inadvertent, because the writer was not careful when taking notes (review pp. 76–80), because he does not understand what plagiarism is, or because he is not conscious of what he is doing.

11.5.1 Plagiarism Defined

You plagiarize when, intentionally or not, you use someone else's words or ideas but fail to credit that person. You plagiarize even when you do credit the author but use his exact words without so indicating with quotation marks or block indentation. You also plagiarize when you use words so close to those in your source, that if you placed your work next to the source, you would see that you could not have written what you did without the source at your elbow. When accused of plagiarism, some writers claim *I must have somehow memorized the passage. When I wrote it, I certainly thought it was my own.* That excuse convinces very few.

> **Intentional Plagiarism is Stealing**
>
> Students who intentionally present the work of others as their own do not always recognize the damage that action does—a matter we address in Part IV. But sometimes they seem not even to know that they are stealing. Colomb once had to arbitrate a dispute between two students who turned in identical papers for the same class. Confronted with the evidence, the first student admitted that she had copied a paper shown to her by the other student. Hearing this, the other student became irate, complaining that the first student had no right to copy his paper, because he had gotten it out of his fraternity's files, and only members of his fraternity had the right to turn in those papers as their own!

11.5.2 Straightforward Plagiarism of Words

When you want to use the exact words you find in a source, stop and think. Then:

- type a quotation mark before and after, or create a block quotation (see the Quick Tip at the end of this chapter),

- record the words *exactly* as they are in the source (if you change anything use square brackets and ellipses to indicate changes), and
- cite the source.

Those are the first three principles of using the words of others: unambiguously indicate where the words of your source begin and end, get the words right (or indicate changes), and cite the source. If you omit the first or last step, intentionally or not, you plagiarize.

11.5.3 Straightforward Plagiarism of Ideas

You also plagiarize when you use someone else's ideas and you do not credit that person. You would be plagiarizing, for example, if you wrote about problems using the concepts from Chapter 4 and you did not credit us, even if you changed our words, calling Conditions, say, *Predicament,* and Costs *Damages.*

If you use someone else's ideas, credit your sources up front. If you write several pages based on the work of another, don't postpone mentioning that fact until a footnote at the end.

A tricky situation arises when you get an idea on your own, but you later discover that someone else thought of it first, or something close to it. In the world of research, priority counts not for everything, but for a lot. If you do not cite that prior source, you risk having people think that you plagiarized it, even though in fact you did not.

An even trickier situation is when you are using ideas that are widely known in your field. Sometimes the idea is so familiar that everyone knows who gets credit for it, and you might be thought naïve if you cited it. You might, for example, mention Crick and Watson when you talk about the helical structure of DNA, but you would probably not cite their article announcing that discovery. At other times, however, the idea seems to you common knowledge, part of the background in your field, and you do not know who first published it. Since you can't track down everything you say in your paper, these are cases in which even the most scrupulous students can misstep. All we can say is, *When in doubt, ask your teacher and give credit where you can.*

11.5.4 Indirect Plagiarism of Words

It is trickier to define plagiarism when you summarize and paraphrase. They are not the same, but they blend so seamlessly that you may not even be aware when you are drifting from summary into paraphrase, then across the line into plagiarism. No matter your intention, close paraphrase may count as plagiarism, *even when you cite the source.*

Another complication is that different fields draw the line in different places. In the law, you are expected to paraphrase statutes and court decisions very closely. In the sciences, writers often cite and then closely paraphrase the main finding of an article, though not other parts. But in fields that use a lot of direct quotation, such as history and English, close paraphrases are risky.

For example, this next paragraph plagiarizes the first paragraph in this section, because it paraphrases it so closely:

> It is harder to describe plagiarism when summary and paraphrase are involved, because while they differ, their boundaries blur, and a writer may not know that she has crossed the boundary from summary to paraphrase and from paraphrase to plagiarism. Regardless of intention, a close paraphrase is plagiarism, even when the source is cited. This paragraph, for instance, would count as plagiarism of that one (Booth, Colomb, and Williams, 169).

The following is borderline plagiarism:

> Because it is difficult to distinguish the border between summary and paraphrase, a writer can drift dangerously close to plagiarism without knowing it, even when the writer cites a source and never meant to plagiarize. Many might consider this paragraph a paraphrase that crosses the line (Booth, Colomb, and Williams, 169).

The words in both these versions track the original so closely that any reader would recognize that the writer could have written them only while *simultaneously* reading the original. Here is a summary of that paragraph, just on the safe side of the border:

> According to Booth, Colomb, and Williams, writers sometimes plagiarize unconsciously because they think they are summariz-

ing, when in fact they are closely paraphrasing, an act that counts as plagiarism, even when done unintentionally and sources are cited (169).

11.5.5 Becoming Aware That You Are Plagiarizing

Here is a simple test for inadvertent plagiarism: Be conscious of where your eyes are as you put words on paper or on a screen. If your eyes are on your source at the same moment your fingers are flying across the keyboard, you risk doing something that weeks, months, even years later could result in your public humiliation. Whenever you use a source extensively, compare your page with the original. If you think someone could run her finger along your sentences and find synonyms or synonymous phrases for words in the original in roughly the same order, try again. You are least likely to plagiarize inadvertently if, as you write, you keep your eyes not on your source but on the screen or on your own page, and you report what your source has to say *after those words have filtered through your own understanding of them.*

11.6 THE LAST STEPS

If you are a sentence-by-sentence drafter and reach the end, you are ready for the last stage. But if you adopt the faster but rougher let-it-flow approach, you will have some cleaning up to do. What you are aiming at is a legible first draft that won't distract you with garbled sentences and multitudes of surface errors. Don't worry about catching them all: you'll clean up more carefully nearer the end.

Go back and fill in the blanks: Type in quotes, add footnotes, do the mechanical work that you skipped. (If you use a word processor, clean up in stages, printing out hard copy for each new stage. If you use a typewriter, collect all your quotes and draft your footnotes, inserting them when you retype your draft.)

Now read through your draft as rapidly as you can, preferably out loud to a friend or roommate. You are looking only to gauge the flow of your argument. If you stumble over a sentence, mark it but keep moving. If paragraphs sound disconnected, add a transition if one comes quickly or mark it for later. If points don't seem to follow, note where you became aware of the problem and move on. Unless you are a compulsive editor, do not bother getting every sentence perfect, every word right. You may be making so many changes down the road that at this stage there is no point wasting time on small matters of style, unless, perhaps, you are revising as a way to help you think more clearly. Once you have a clean copy with the problems flagged, you have a revisable draft.

At this point, however, you face a problem that vexes every writer: you must determine whether your report will make sense to your readers. You must now try to read it *through their eyes,* imagining how *they* will understand it, what *they* will object to, what *they* need to know early so that they can understand something later. Some writers fear that this last step compromises their intellectual integrity as the lone pioneer breaking new ground. Certain they have discovered Something Important, they want to believe that the truth of their discovery should speak for itself, needing no clever rhetoric. It is the Truth-as-Hero story, an anti-rhetorical position articulated 2,500 years ago by Socrates, and debated ever since.

Despite that Platonic ideal of unadorned truth, knowledge is never just discovered, presented, and accepted. New ideas are always created and then *shaped* by writers who anticipate the needs, beliefs, and objections of their readers. By imagining themselves in a conversation with them, wondering what they think, what they must understand, writers discover better what they themselves *can* think. The best means to that end is careful revision.

Perhaps the biggest difference between experienced writers and beginners is their attitude toward this first draft. The experienced writer takes it as a challenge: *I have the sketch, now comes the hard but gratifying work of discovering what I can make of it.* The beginner takes it as a triumph: *Done! I'll change that word, fix this comma, run the spell checker, and <Print>!* A first draft is indeed a victory, but resist that easy way out. In the remaining chapters, we will describe ways to revise your drafts not as a chore but as a way to keep the creative juices flowing.

QUICK TIP:
USING QUOTATION AND PARAPHRASE

Regardless of your field, you have to rely on the research of others and report what they have found. But the practices of your field will determine how you do that.

HOW TO QUOTE AND PARAPHRASE

In the sciences and some social sciences, researchers rarely quote sources directly; instead they paraphrase and cite them. The process is simple: In your own words, restate either the finding or the data that you want to use. Then be sure to cite the source using the form appropriate to your field. Make the name of the source a direct part of your own sentence only if the source is an important one and you want to call attention to it.

> A number of processes have been suggested as causes of the associative-priming effect. For instance, in their seminal study Meyer and Schvaneveldt (1971, p. 232) suggested two, namely *automatic (attention-free) spreading activation* in long-term memory and *location-shifting*. Neely (1976) similarly distinguished between a process of automatic-spreading activation in memory and a process that depletes the resources of the attentional mechanism. More recently, a further associative-priming process has been studied (de Groot, 1984).

The writer thought that Meyer, Schvaneveldt, and Neely were important enough to name in her sentences, but cited de Groot only as a minor reference.

In the humanities and some social sciences, researchers sometimes paraphrase sources, but are more likely to quote them. You have three options.

• Introduce a quotation with a colon or an introductory phrase:

> Plumb describes the Walpole administration in terms that remind one of the patronage system in U.S. cities: "Sir Robert was the first English politician to understand how to use the loyalty of people whose only qualification was his sponsorship" (p. 343).

> Plumb describes the Walpole administration in terms that remind one of the patronage system in U.S. cities. He claims that "Sir Robert was the first English politician to understand how to use . . . ."

- Weave the quotation into your own sentence (but be sure that the grammar of your part of the sentence matches the grammar of the quotation):

> Plumb speaks in terms that remind one of the patronage system in modern U.S. cities when he describes how Walpole was able "to use the loyalty of people whose only qualification . . . ."

> Jameson was never comfortable with the decisions of the Tribunal, and he often "complain[ed] . . . that something had to be changed" (1984, p. 44).

[Note that when this writer changed the original, she used square brackets and ellipses to indicate every change.]

- Set off in a "block quote" quotations of three or more lines. When you use a block quote, be sure that the quotation connects to what has gone before, and that just before or just after the quote you make clear why you are quoting it.

> After the Restoration in 1660, English philosophers and moralists continued to complain that people were motivated by money and material goods, which was, of course, nothing new. But these thinkers did believe they saw a change: a new form of "mercenary virtue" that tried to offer material incentives for good behavior. These new complaints culminated in the work of Shaftesbury:
>
>> Men have not been contented to show the natural advantage of honesty and virtue. They have rather lessened these, the better, as they thought, to advance another foundation. They have made virtue so mercenary a thing, and have talked so much of its rewards, that one can hardly tell what there is in it, after all, which can be worth rewarding. (p. 135)

- Do not begin a sentence with quoted material and end it with your own words. Begin your sentences with your own words and end them with the quoted material.

WHEN TO QUOTE AND PARAPHRASE

No matter what your field, you have to learn how much to rely on the work of others. If you quote or cite others too often, you will seem to offer too little of your own work; quote too little, and readers may think your claims lack support or may not see how your work relates to that of others. We can't give you definitive rules for deciding when and how much to quote or paraphrase, but there are some rules of thumb.

Use direct quotations:
- when you use the work of others as primary data,
- when you want to appeal to their authority,
- when the specific words of your source matter because
  - those words have been important to other researchers,
  - you want to focus on how your source says things,
  - the words of the source are especially vivid or significant,
  - you dispute your source and you want to state her case fairly.

Paraphrase your sources:
- when you are more interested in content, in findings or claims, than in how a source expresses himself,
- when you could have said the same thing yourself more clearly.

Don't quote simply because it's easier or because you think you don't have the standing to speak for your sources. Keep your quotations as short as possible, and under no circumstances cobble together a paper out of a series of quotations. You must make your own argument with your own claims and evidence.

CHAPTER TWELVE

Communicating Evidence Visually

This chapter discusses matters that most writers think about, if at all, only at the very end of the writing process. But, depending on your field, you should think about presenting your evidence visually in the first stages of drafting.

READERS WILL JUDGE the quality of your research by the significance of your claim and the power of your argument. But before they can make that judgment, they must understand what you have written. To that end, in Chapters 13 and 14 we discuss how to create a report that is coherently organized, written in a readable prose style. But if your data consist of discrete items—numbers; lists of names, places, objects; or even concepts legitimately reduced to a few words—you can often help your readers understand that data and thus your argument in another way: visually, through tables, charts, graphs, diagrams, maps, and visual signals of logical structure.

12.1 VISUAL OR VERBAL?

Whether you choose to present data visually or verbally depends on

- the kind of data,
- how your readers can best understand them,
- how you want your readers to respond to them.

You communicate best in words when your information is qualitative and not easily presented formally, or when your readers are strongly "word" oriented, as most are in the humanities. With other readers, however, you can communicate more effectively with tables, graphs, or charts, if your data have these characteristics:

- They include independent elements. These can be discrete elements that are well defined and stable, called "cases"—persons, places, things, or concepts. Or the independent element can be an "independent variable,"

a measurement scale that does not change in response
to other variables—time, temperature, distance, and
so on.

• The independent elements are systematically related to
quantities or qualities, called "dependent" variables,
data that do change in response to external causes.

For example, the next two paragraphs have three independent
elements (the three counties) and many dependent variables. But
only in the second are the elements and variables related systematically enough to present visually:

> The populations of Oswego, Will, and Tuttle counties fell
> as a result of a 31.6% drop in manufacturing from 1970 through
> 1990, and a 65.9% drop in family farms in Tuttle alone, starting
> in 1980, when farming there employed more than 55% of the
> work force, through 1990, when it employed less than 30%. As
> employment fell, so did the numbers of those moving into Os-
> wego and Tuttle by 73%.

> As a result, from 1983 through 1993, the population of
> these counties steadily declined: Tuttle by 10,102, or 49.3%,
> from 20,502 to 10,400; Will from 16,651 to 15,242, or 8.5%;
> Oswego 39.1%, from 15,792 to 9,614, for a net loss of 6,178.
> The differences can be accounted for by the fact that Tuttle and
> Oswego depend on agriculture, Will mainly on small manufac-
> turing.

In the first paragraph, we cannot systematically align the counties
with the dependent variables in a way that shows the complex
causal relationships that the paragraph communicates. Only words
will serve. In the second paragraph, the counties systematically
correlate with data about industry, population, and changes. Those
relationships would be more easily seen in a table:

Table 12.1: Population Decreases by County, 1983–1993

County	Industry	1983	1993	Decrease	%
Tuttle	Farming	20,502	10,400	10,102	− 49.3%
Oswego	Farming	15,792	9,614	6,178	− 39.1%
Will	Mnfctrng	16,651	15,242	1,409	− 8.5%

To communicate these data with more rhetorical impact, we could use a bar chart that invites us to "see" an image of these differences and to compare them. Notice that the bar chart communicates fewer data, less precisely. (We label graphs and charts "figures.")

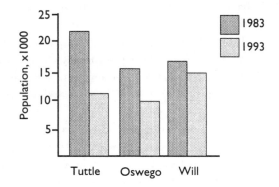

Figure 12.1: Population Decreases by County: 1983–1993

Finally, we could present these data still more strikingly with a graph, in a way that makes us see these changes as a story:

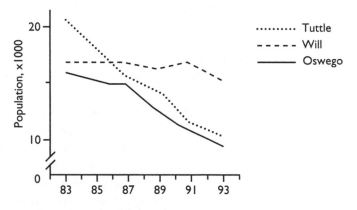

Figure 12.2: Population Decreases by County: 1983–1993

In this chapter, we discuss when to use and construct tables, charts, graphs, and other illustrations so that your readers can grasp complex information easily while feeling the rhetorical impact that you intend.

12.2 SOME GENERAL PRINCIPLES OF CONSTRUCTION

As with everything else in your project, spend a few minutes *planning* what you want your table, chart, or graph to accomplish.

1. How *precise* will your readers want the data to be?
   - Tables are more precise than charts and graphs.
2. What kind of rhetorical and visual *impact* do you want readers to feel?
   - Tables seem to present data objectively. Although you do select the data, those data seem not to reflect your own interpretation. Present data in tables if you want to be precisely descriptive and reduce rhetorical impact.
   - Charts and graphs are visually more striking. They encourage readers to react to their visual image.
   - Charts invite readers to make comparisons.
   - Graphs invite readers to see a story.
3. Do you want your readers to see a *point* in the data?
   - Tables encourage readers to interpret the data.
   - Charts and graphs seem to make your point more directly.

Regardless of which form you choose, your readers will understand your data most easily if you follow three principles of construction.

1. The more organized, the better. Arrange the elements by a principle that reflects how you want your readers to *use* the table or figure:
   - Order independent elements by a principle that reflects whatever patterns you want readers to notice.
   - In tables, organize data so that readers' eyes are drawn to the elements that you most want them to notice.
   - In charts, if possible order the bars so that they form a coherent shape that you intend: a slope upward or downward, a bell curve, a level line, etc.
   - In graphs, if possible arrange the variables so that the slope implies a story that supports your point.
2. The simpler, the better.
   - Limit the cases—names of persons, places, things— to four for a graph, six or seven for a chart. Use more

than one chart or graph rather than cram a mass of data into one.

- Keep explanatory words on the chart or graph to a minimum.
- Use only a few coordinated type fonts. Avoid using just upper-case letters.
- In charts and graphs, keep visual contrasts simple: black, white, and one or two shades of gray—avoid cross-hatching, stripes, etc.

3. Most important, just before or just after the reader sees those data, state the point that you think they make and that you want your reader to understand. Point out those differences, similarities, anomalies, or patterns that you think most significant. If the data hold no surprises, say so.

A Word of Caution

Most of you will create your visuals on a computer, using software that generates charts and graphs automatically. Beware, however: most software packages create visuals that look good but do not communicate as well as they should. Software developers are more interested in glitzy pictures, the fancier the better, than in visuals that tell their story effectively. If you use charting software, resist the temptation to use all of its features. Avoid options that depart from the principles you find here. Expect that you will have to import the visual created by your charting software into a graphics package in order to adjust it according to our principles.

12.3 TABLES

Tables are useful when you want to convey precise values, when you have to present a large array of data, or when you do not know (or do not want to say) which aspects of the data will be most important to readers who need the data in front of them so that you can refer them to the items. Tables seem objective and encourage readers to draw their own conclusions. There are two types, number tables and word tables.

12.3.1 Number Tables

The first principle in constructing number tables is to help readers see what you want them to. If readers will use a table not to compare values but to find specific values that you cannot predict, arrange the items in some default order: in this case, Table 12.2 arranges municipalities alphabetically, revenues from whole to parts.

Table 12.2: Revenues for Selected Municipalities (in millions)

| | | Total | Sales Taxes | | Property Taxes | User Fees |
| | Status | $ | State | City | Taxes | User Fees |
	Status	$	$ %	$ %	$ %	$ %
Alameda	Tnshp	1.43	0.26 (18)	0.00	0.97 (68)	0.20 (14)
Blythe	City	7.18	2.37 (33)	2.37 (33)	2.44 (34)	0.00
Capital	City	20.02	4.00 (20)	7.41 (37)	7.41 (37)	2.60 (13)
Danberg	Tnshp	3.03	1.15 (38)	0.00	1.48 (49)	0.39 (13)
Eden	Vllge	10.32	1.55 (15)	0.00	5.16 (50)	3.61 (35)

If on the other hand you want readers to see specific *differences*—in this case, that cities with a sales tax depend less on property taxes—then the salient *comparisons* should be ordered top to bottom, even highlighted.

Table 12.3: Revenues for Selected Municipalities (in millions)

| | | Property Taxes | Sales Taxes | | User Fees | Total |
| | Status | Taxes | City | State | User Fees | Total |
	Status	$ %	$ %	$ %	%	$
Alameda	Tnshp	**0.97 (68)**	**00 (00)**	0.26 (18)	0.20 (14)	1.43
Eden	Vllge	**5.16 (50)**	**00 (00)**	1.55 (15)	3.61 (35)	10.32
Danberg	Tnshp	**1.48 (49)**	**00 (00)**	1.15 (38)	0.39 (13)	3.03
Capital	City	**7.41 (37)**	**7.41 (37)**	4.00 (20)	2.60 (13)	20.02
Blythe	City	**2.44 (34)**	**2.37 (33)**	2.37 (33)	0.00	7.18

When readers see values that are grouped, they can mentally add and subtract as their eye travels down and then compare changing values most easily.

Some additional principles:

1. List and name the independent elements on the left vertical axis. Keep in mind that readers generally take what is at the left as the cause or source of what appears to their right.

2. List dependent variables in columns, left to right, labeled at the top.

3. If it makes sense, provide an average or median at the bottom of the table so that readers can gauge the range of variation.

4. If you are concerned more with making a point than providing precise data, round your numbers so that readers can compute values from only the first two (or at most three) digits.

5. If a table has more than seven rows, put additional space between every fourth or fifth row.

Remember to *interpret* the table for your reader in your text. Don't just repeat in words what the table presents in numbers.

12.3.2 WORD TABLES

Word tables must express dependent variables concisely.

Table 12.4: Key Features of Tables, Charts, and Graphs

	Precision	Rhetorical Impact	Implied Form
Tables	high	objective	descriptive
Charts	low	objective/subjective	descriptive/narrative
Graphs	low	subjective	narrative

The risk with word tables is that they seem reductive, leading readers to feel that you have oversimplified concepts and eliminated nuance. So use word tables only for conceptual relationships that are straightforward and unnuanced. Most readers would dismiss Table 12.5 as crude overgeneralization:

Table 12.5: Periods in European Culture

Period	Religious Belief	Desire for Order	Individualism
Medieval	very high	high	low
Renaissance	high	medium	medium
Enlightenment	medium	very high	high
Modern	low	very high	high
Post-modern	low	low	low

12.4 CHARTS

Charts help readers understand generally (not precisely) how several independent cases or categories vary by one or just a few dependent variables. They give readers an image of the data:

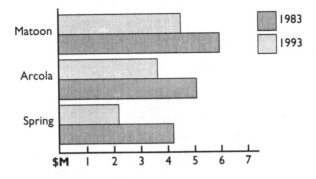

Figure 12.3: Increase in Municipal Revenue, 1983–1993

Charts are descriptive, but they can imply a story if you arrange the data so that they seem to change systematically, even though they don't:

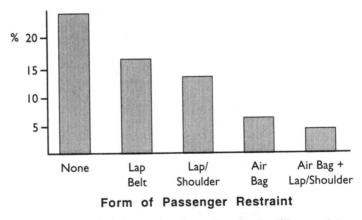

Figure 12.4: Collisions with at Least One Fatality (30 + m.p.h.)

As readers move left to right, they seem to *see* fatalities decline as protection increases, indicating a hopeful trend for readers concerned about auto safety. But if the writer wanted to shake up readers complacent about safety, the chart would seem to tell its story better with the order reversed, the bars "rising" toward higher death rates.

12.4.1 Bar Charts

Effective bar charts follow a few principles:
1. If you arrange bars horizontally (as in Figure 12.3),
   - list independent elements at the left, top to bottom;
   - arrange the dependent variable at the bottom, left to right.
2. If you arrange bars vertically (as in Figure 12.4),
   - list independent elements along the bottom, left to right;
   - arrange the dependent variable along the left, bottom to top.
3. If you want to communicate specific values, insert numbers in or at the end of each bar.
4. Avoid three-dimensional bars. Readers have to interpret whether their volume or length is the salient image. Especially difficult are charts whose "bars" are pyramids, cylinders, or complicated iconic shapes.
5. Avoid charts with divided or "stacked" bars. Instead use separate, parallel charts, one for each category.
   - Stacked bars force readers to estimate proportions by eye. In Figure 12.5 who has the largest share of the 35–45 market?

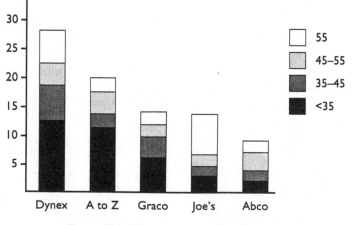

Figure 12.5: Market Share by Age Segment

- Stacked bars also force readers to calculate proportions of proportions. In Figure 12.5 what proportion of the whole market is over 45?

6. If you insist on using stacked bars, help your readers by following these principles:
   - Arrange the segments in some principled order, bottom to top.
   - Use the darkest or most saturated colors at the bottom, lightest at the top. Remember that readers tend to over-estimate the magnitude and importance of darker sections.
   - Use numbers and connecting lines to clarify proportions.

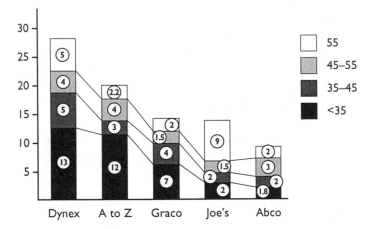

Figure 12.6: Major Competitors' Market Share by Age Segment

You can also use a *point* chart, which does the same thing as a bar chart but is less busy. Here are some of the same data as in Figure 12.6, presented as parallel dot charts. (When creating parallel charts, be sure to use the same scale.)

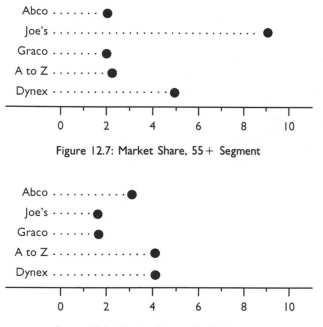

Figure 12.7: Market Share, 55+ Segment

Figure 12.8: Market Share, 45–55 Segment

If we did this with the other age categories, readers could see more clearly and quickly how these competitors control different markets.

12.4.2 Pie Charts

Pie charts are favorites of newspapers and annual business reports but are rarely useful. At best, they allow readers to see crude proportions among a *few* elements that constitute 100% of a whole. They are hard to read when they have more than four or five segments, particularly when the segments are thin. They are especially cumbersome when readers have to look at a key to match the patterns in the segments with categories. Compare how much easier it is to interpret the same information on a bar chart than on a pie chart:

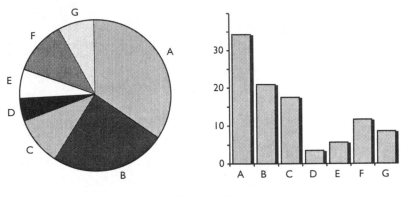

Figure 12.9

1. Avoid pie charts. If you insist on using them, do so *only*
   when your readers need to see just a few imprecise comparisons and the differences are unmistakable at a glance.
2. Arrange the segments in an order meaningful to your readers, beginning at 12 o'clock and moving clockwise. If you
   have no better order, arrange the segments from largest to
   smallest.
3. If one segment is significant, emphasize it.
   * Make the emphasized segment the darkest or most satu-
     rated color, adjacent shades as contrastive as possible.
   * For special emphasis, break that segment out from the
     rest.

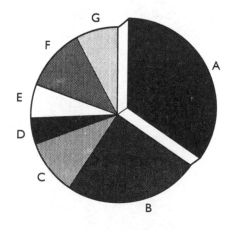

Figure 12.10

Other Volume Charts are also favorites of tabloids but have no place in academic reports. They have the limitations of pie charts and are harder to judge by eye. Create a chart like this and experienced researchers will think you somewhat silly:

Figure 12.11: Oil Imports, 1980–90

12.5 GRAPHS

Graphs do not easily communicate precise values, but they can effectively show rough relationships among many points.

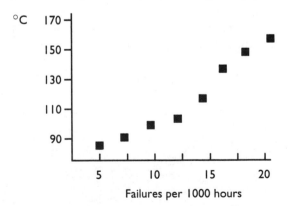

Figure 12.12: Failures Increase with Operating Temperature

Graphs are especially effective at presenting an image of data that move continuously along a line:

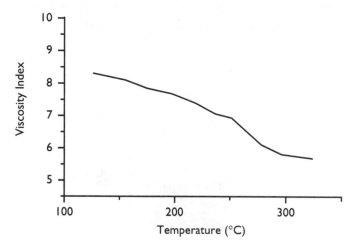

Figure 12.13: Viscosity Decreases as Temperature Increases

Be aware that readers interpret graphs as a story about some entity changing through time, and that readers will project the trend off the chart.

If you have several independent cases, use separate graphs. Keep the number of lines on graphs few, and make the contrast among them strong. Readers have difficulty following more than three lines, especially when they cross, as in 12.14.

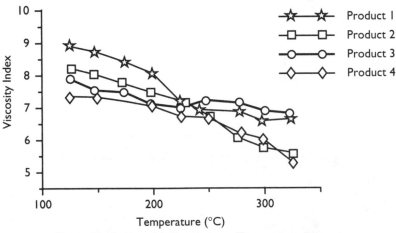

Figure 12.14: Viscosity Decreases as Temperature Increases

If you plot two or three lines that represent portions of a total, you can create an "area" plot by filling the spaces between lines with color or shades of gray. Put the largest quantity on the bottom and fill it with the darkest color. Then order the rest from larger to smaller, with increasingly lighter colors.

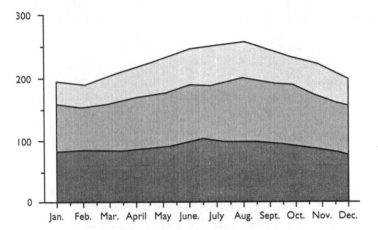

Figure 12.15: Airborne Particulates from Public Incinerators (parts per million)

12.6 CONTROLLING THE RHETORICAL IMPACT OF A VISUAL

Generally, the kind of data should determine the kind of visual. But consider as well the rhetorical impact you want to communicate. For example, Figure 12.16 shows profits for two products over thirteen years.

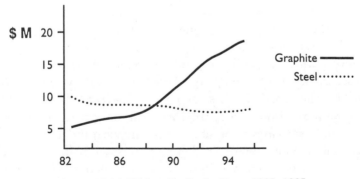

Figure 12.16: Widget Profits by Type, 1982–1995

Such a line graph is the default way to present data about how two independent cases (product types) change across one independent variable (time) and one dependent variable (profit levels). It emphasizes the different movements of the two products, showing readers that graphite is more profitable.

You can, however, tell an apparently different story with the same data, if you present them not in a line graph but in an area plot:

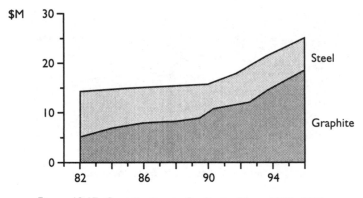

Figure 12.17: Contribution to Profits by Type, 1982–1995

The data here are identical to Figure 12.16, presented with the same accuracy and precision. A reader experienced with both types of graphs could derive the same information from 12.17 as from 12.16, though only with some difficulty.

Notice, however, how different is the impact of the image. In Figure 12.16 the line for steel profits declines, but in 12.17 it seems to rise sharply. The *area* signaling steel profits shrinks over time, but the *image* is of a line rising. In 12.16, we see the image of a company with one good product and one less good one. In 12.17, we see the image of a company whose total profits have been rising steadily. The image of this same data is different still in 12.18.

You must also consider the rhetorical differences in communicating different kinds of numbers, not only what the numbers measure (sales in units, total sales dollars, profits, etc.) but also whether the numbers represent absolute values ("raw numbers") or relative

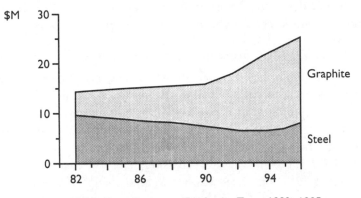

Figure 12.18: Contribution to Profits by Type, 1982–1995

values (percentages, proportions, etc.). In each of the graphs concerning Widget profits, the dependent variable is profits in millions of dollars. Those same data might also be communicated not as raw numbers but as proportions, changing the visual impact yet again.

Compare 12.16 with 12.19, which is based on the same data, but now as a *proportion* of total profits, which climb steadily from 1982 to 1995. This makes steel widgets seem even worse than do the raw numbers in 12.16.

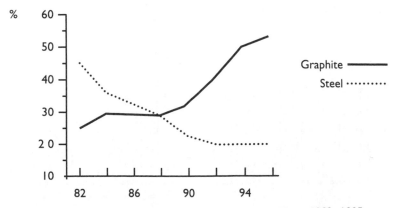

Figure 12.19: Contribution to Total Profits by Type, 1982–1995

If you decide to depart from the default visual form for your kind of data, be certain you have a good reason to do so.

12.7 Visual Communication and Ethics

When you select a visual for its impact, remember that your rhetorical decision has an ethical dimension. Suppose, for example, you are presenting data about Widget profits to answer a question about how Widget, Inc., is doing in general. In that case, any of the five graphs would be appropriate. But if the question concerned the future of the steel division, a reader might reasonably think that 12.17 or 12.18 was less helpful than 12.16 or 12.19. In fact, readers might think that 12.17 was so misleading as to be deliberately so.

Whenever you present data visually, you have to choose between telling your story to get the right impact and your responsibility not just to the facts but to their *appearance*. Because tables, charts, and graphs seem objective, inexperienced readers can be fooled by them, but experienced readers will become suspicious if they think you are distorting your images to serve your story. Unfortunately, it is sometimes difficult to distinguish effective rhetorical impact from unfair manipulation. This tricky decision applies to everything in your report, but it is especially important with visual devices, because they seem to present data so clearly and so powerfully.

For example, compare the two charts in Figure 12.20. The data in the two are identical, but look at the slope of the bars.

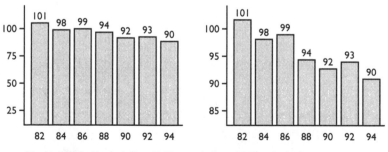

Figure 12.20: Capital City Pollution Index, 1982–1994 (July averages)

On the left, the slope represents changes in data points more accurately, because the scale begins at 0. On the right, the slope is much sharper, because the scale begins at 80: consequently, the bar for 1994 is about half the size of the one for 1982, *even though the difference in absolute values is only 10%*. As a result, the chart on the right suggests more improvement than the one on the left, a story that might mislead some readers and that others would consider dishonest.

The question of honesty in 12.20 is mitigated by the fact that the bars are clearly labeled with precise values. But a writer who truncates the vertical axis of a graph to make a slope seem sharper may cross the line of honesty, because to the viewer the *slope* of a graph is always the predominant image. Simply by changing the scale on a vertical index, you can communicate what seem to be different stories:

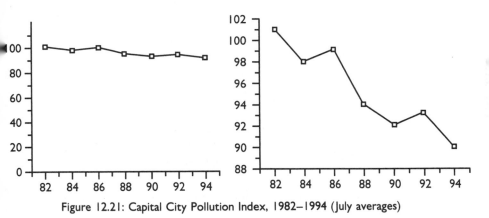

Figure 12.21: Capital City Pollution Index, 1982–1994 (July averages)

On the other hand, it is not always easy to distinguish what is "objective" from what is "ethical." Suppose you are an environmental scientist and you know that any expert would consider these seemingly small decreases to be highly significant. But you are certain that your statistically unsophisticated but influential readers will dismiss the differences as meaningless because the *visual* differences in the graph on the left are so slight. If you are certain that large visual differences would better communicate the actual scientific *significance* of those differences, then the question which graph is more honest becomes less clear.

12.8 CONNECTING WORDS AND PICTURES

This chapter has focused on visuals, but visuals are only one element in a text mostly of words. Visuals can't speak for themselves. You must tie your words and pictures together.

1. Always label visuals clearly.
   - Put a caption on each table, chart, and graph. (Many publications put the caption under its visual, but when a caption is placed above a visual, readers are encouraged to read it first, which helps them to know what to look for.)
   - If possible, the caption should suggest the point of the visual. At the least, the caption must indicate the kind of data presented.
   - Label each axis, including units of measure.
   - Label each plot when there are more than one.
2. Number tables and figures separately.
3. Locate tables and figures as close as possible to the text that discusses them.
4. Always refer to tables and figures in the body of your text. Tell readers what to see, and if you want readers to take away a point from a table or figure, say so explicitly.
5. If necessary, highlight the portion of the image that is important.

12.9 SCIENTIFIC VISUALIZATION

In this chapter, we have discussed visuals with few variables and data points. But some areas of science work with thousands, even millions of data points, data sets so complex that we can comprehend them only through what is called "scientific visualization." Unless you are an advanced researcher, you are unlikely to require such elaborate visualization techniques. And even if you do, the process will be largely done by computer software. You will face the same rhetorical considerations, but your control over the process will depend on the evolving state of software and on your expertise in not just using that software but in understanding its rhetorical potential.

12.10 ILLUSTRATIONS

Tables, charts, and graphs are not the only kinds of visual tools. Researchers also use visuals to illustrate conceptual matters. Except for the examples in this chapter, we have used no tables or graphs in this book, but we have used several diagrams. We cannot go into detail about how to construct these other visuals, but here are a few of the common forms used in a variety of fields of study.

To illustrate this .... use this.	
process	flow chart
	decision tree
logical relationships	diagram
	matrix
object	line drawing
	drawing
	photo
parts of a complex object	line drawing
	exploded view
action/step in a process	line drawing
	drawing
	photo
spatial relationships	line drawing
	drawing
complex detail	photo
	drawing
research settings	photo
	diagram

12.11 Making the Logic of Your Organization Visible

In some fields—particularly the humanities—writers use few visual resources to signal their logic. They may deploy an occasional heading, sneak in an extra space between sections, emphasize a few words with italics and boldface, but infrequently. For the most part, they rely on the intrinsic clarity of their organization and prose style to communicate the logic of their argument. In fact, some claim that to do otherwise panders to semiliterate readers who can't read well enough to understand even moderately complex ideas.

But in most other academic areas and in almost all nonacademic ones, writers freely use visual resources when those devices will help readers better understand the logical structure of information. They unhesitatingly decompose sentences and paragraphs into tabular form, not just to signal the structure of what they are communicating but to relieve the grayness of solid type. In this book, we have presented information in a tabular way at every opportunity. Compare this next paragraph with what you read on p. 178:

> There are some general principles of construction. As with everything else in your project, spend a few minutes *planning* what you want your table, chart, or graph to accomplish. How *precise* will your readers want the data to be? Tables are more precise than charts and graphs. What kind of rhetorical and visual *impact* do you want readers to feel? Tables seem to present data objectively. Although you do select the data, those data seem not to reflect your own interpretation. Present data in tables if you want to be precisely descriptive and reduce rhetorical impact. Charts and graphs are visually more striking. They encourage readers to react to a visual image. Charts invite readers to make comparisons. Graphs invite readers to see a story. Do you want your readers to see a *point* in the data? Tables encourage readers to interpret the data. Charts and graphs seem to make your point more directly.

Now in fact, some readers will claim to prefer this to bullets and headings, because they believe that they can absorb it better—

especially readers in the humanities. But if we can trust what research says about how most of us read and understand, you should
assume that most readers prefer to see information structured visually and, when they do, will absorb more of it, understand it better,
and retain it longer.

12.12 Using Visual Forms as an Aid to Thinking

These visual devices help you communicate complex data, but
they have another important use: They can also help you *discover*
patterns and relationships that you might otherwise have missed.

Before you draft, try playing with your data visually. Spend
time arranging and rearranging your information in different forms
and in different ways—in a graph, chart, table, or diagram. You
might not actually include them in your final draft, but visuals can
stimulate your thinking and help you organize your ideas. The more
ways you can structure and restructure your data, particularly if it
forces you to step out of the ruts of your ordinary thinking, the
better you will understand those data and the more chance you
have to discover things that may surprise you. What would a graph
look like that contrasted Macbeth's moral development with Lady
Macbeth's? What would be the dependent variables?

These visual devices can even suggest ways to organize your
report. For example, you might not actually present the word table
we used on p. 181, but its categories on the horizontal and vertical
axes suggest different ways to organize your materials—by period
or by the categories of belief, order, and individualism.

When you have a draft, try breaking a paragraph or even a
section that feels long and garrulous into the bulleted and sub-
pointed tabular arrangement we've used here. If you cannot even
begin to do so, you may have a problem with your organization—
your sentences may just string out, one after the other, in no order
other than *well, here's one more idea.* Only when you have arranged
your prose in a coherent, organized way can you even begin to
think about using bulleted subpoints.

Use headings liberally (but see pp. 208–9). They help your
readers identify where one section stops and another begins, but
headings can also help *you* diagnose your own organization. If you
cannot decide where to put a heading or what words should go

into it, you may have a problem, and if *you* have a problem, your reader will too.

Like other formal devices, visuals encourage you to discover ideas and relationships that you might not have seen otherwise. In the next three chapters, we will discuss other rhetorical forms that can also stimulate your thinking and improve your understanding of your project from its beginning to its end.

Quick Tip:
A Consumer's Guide to Visiting a Writing Tutor

Many colleges provide tutors to help students with their writing. (If you don't know how to find one, ask your English department or dean of students. If they look at you blankly, urge them to provide such help.) Tutors can help when you are struggling with a paper, but they can't think or write for you or help you if you don't know how to use them. Here's how.

If possible, find a tutor who knows something about your subject matter. You have seen how thinking and writing are intertwined. Though tutors are trained to deal with different kinds of papers, you'll get better advice if the tutor understands your field.

Plan. Before you see a tutor, be sure that you can describe what you have done, what not, and what parts of the task trouble you. The clearer you can be, the better advice you will get.

Some tutoring programs require students to submit drafts or outlines before meeting with a tutor. Follow that procedure, even if your tutors do not. At least prepare the materials your tutor will need in order to help you.

First, prepare an outline that shows the tutor where your paper stands. A sentence outline that lists main points is better than a topic outline, but any outline is better than none. It should show which parts you have drafted, which you are relatively sure of, and which are only guesses. If you are at the earliest stages of research and cannot formulate an outline, sketch your specific topic, either in a paragraph or two or as a list of topics you have begun to investigate.

Next, if you have a draft, prepare two clean, double-spaced copies. One should remain clean, ready for the tutor to mark up. The other you should mark up as follows:

1. Draw a line between the introduction and the body of the paper and another between the end of the body and the conclusion. If the body is long enough to divide into two- or three-page-sized sections, put lines there as well.

2. Highlight the main point of your paper. If you have di-

vided the paper into sections, highlight the main point of each section.

3. Circle the words near the end of the introduction that name the key concepts you will develop as themes in the rest of the paper. Then circle those words and words similar to them throughout.

4. If you have divided your paper into sections of three pages or longer, repeat steps 2 and 3 for each section.

5. Add headings for each major section, even if you intend to remove them after the tutorial session.

6. Mark in the margins any problem areas where drafting was particularly difficult or where you are dissatisfied with what you've done.

Be sure to take your assignment sheet and anything else your instructor gave you in writing.

Before you leave, get a plan of action *in writing*. Many students discover that while they are talking to a tutor, they think they understand what to do next, but that the plan evaporates a few hours later when they sit down to work. Before you leave the tutor, get a written plan of specific ways to improve your paper. If the tutor does not recommend specific actions, ask. You have the right to a plan that you can understand and follow.

CHAPTER THIRTEEN

Revising Your Organization and Argument

What follows may at first reading seem complicated. But if you focus on each step, one at a time, you will find it fairly simple. It will help you analyze your paper more easily and more thoroughly than by just reading and wondering whether it all hangs together.

THE KEY TO REVISING YOUR REPORT is to gauge how it seems not to you but to your reader. To do that, you cannot read it sentence-by-sentence, straight through from beginning to end, thinking to yourself, *Hmmm, maybe this word needs changing, that sentence shortened, but, hey, it all looks pretty good to me.* Revision is a task that requires a level of planning and discipline more deliberate than that.

13.1 THINKING LIKE A READER

First, readers do not read sentence-by-sentence, accumulating information as they go, as if they were collecting beads from a string. They need a sense of structure and, most important, an idea of why they should read your report in the first place. In this chapter, we'll discuss how to diagnose and revise your organization and argument. In the next, we'll discuss style, and in Chapter 15, how to create an introduction that "sells" your readers on the value of your project.

Since readers read each sentence in light of how they see it contributing to the whole, it makes sense to diagnose first the largest elements, then focus on the clarity of your sentences, and only last on matters of correctness, spelling, and punctuation.

In reality, of course, no one revises so neatly. All of us revise as we go, correcting spelling at the same time that we rearrange our argument, deciding to restructure a paragraph as we fiddle with a comma or semicolon. But it is useful to keep in mind that when you revise from the top down, from global structure to sections to paragraphs to sentences to words, you are more likely to discover useful revisions than if you start at the bottom with words and sentences and then work up.

Second, regardless of how you revise, you face a problem that all writers share: you cannot experience your own prose as your readers will, because you know too much about it. When you come to a passage that your readers might stumble over, you sail right through, because you aren't actually reading it; you are only re-minding yourself of what you were thinking when you wrote it.

To help you overcome the problem of your intractable subjec-tivity, we are going to suggest some formal, even mechanical ways for you to analyze, diagnose, and revise your draft, ways that will help you sidestep your too-easy understanding (and too-ready ad-miration) of your own words.

These revisions take time, so start early. Moreover, in the course of revising, you will almost certainly discover something new about your project, some fact or idea that you want to add, some part of your argument that you want to rework. You may think the end is in sight, but revision is as important as any other stage in your project, so don't rush it. In fact, this final stage is when you will come to understand your project most completely.

13.2 ANALYZING AND REVISING YOUR ORGANIZATION

The process consists of four steps:

1. Identify the outer frame of your paper: your introduction and conclusion, and a sentence in each that states your main claim, the solution to your problem. We'll call these your main *point sentences.*
2. Identify the major sections of the body of your paper, their introductions, and the point sentences for each of those sections.
3. Identify in the introduction to the whole paper your central thematic concepts, then track them through the rest of the paper. Then do the same for each section.
4. Step back to grasp the overall shape of your paper.

13.2.1 Step 1: Identify Your Outer Frame and Main Points

Your reader must know three things unambiguously:

- where your introduction ends and the body of your paper begins,

- where the body of your paper ends and your conclusion begins,
- which sentence is the main point sentence in your introduction, and which in your conclusion.

To make those elements absolutely clear, do the following:

1. Start a new paragraph after your introduction and another new paragraph at your conclusion. In fact, put an extra space between your introduction and body and again between body and conclusion. Unless your field disapproves of headings, you'll put headings at these joints to make sure your reader can't miss them.
2. In your introduction, underline the sentence that comes closest to stating your main claim or that points the reader toward it. Ordinarily, that sentence will be the last sentence of your introduction. (Do not consider as a candidate a topic-announcing sentence like *This paper will discuss* . . . . See pp. 93–96.)
3. In your conclusion, do the same thing: underline the sentence that best captures the main point of your paper, your main claim, the sentence that expresses the gist of the solution to your problem.

Now compare the point in your introduction and the point in your conclusion. They should at least not contradict each other. If one is more specific and contestable, it should be the one in your conclusion. If the point sentence in your introduction is vague, unspecific, merely a "topic-announcing" sentence, revise it.

For example, the following introduction and conclusion show what you can do when you apply these tests (we'll assume we have already identified where the introduction stops and the conclusion begins). The introductory paragraph:

> In the eleventh century, the Roman Catholic Church initiated several Crusades to recapture the Holy Land. Two popes called for armies to support this endeavor. In a letter to King Henry IV in the year 1074, Gregory VII urged a Crusade but failed to carry it out. In 1095, his successor, Pope Urban II, gave a speech at the Council of Clermont in which he also called for a Crusade, and in the next year, in 1096, he successfully initiated the First Crusade. I will discuss the reasons that these popes gave for a Crusade.

And the concluding paragraph:

> As we can see from these documents, Popes Urban II and Gregory VII urged the Crusades as a way not just to restore the Holy Land to Christian rule, but also to preserve the political unity of the Church and Western Europe. Urban wanted to conquer the Muslims, but no less importantly to reinforce his authority and control fighting among Europeans by directing their energies elsewhere. Gregory wished to unify the Roman and Greek Churches, but also to prevent the breakup of the Catholic Church and even the Empire. To achieve their political ends, each pope tried to unite people in a common religious fight against the East to prevent them from fighting among themselves and to unify an increasingly divided Church. Thus the Crusades were not just a religious effort to recapture the Holy Land and to save God's faith, as is so widely believed in popular memory, but also a political effort to unify the Church and Europe and save them from internal forces threatening to tear them apart.

The point sentence of the introduction appears to be the last one:

> I will discuss the reasons that these popes gave for a Crusade.

But that sentence is so insubstantial, so vague, so uncontestable, that it does nothing more than announce *I'm going to tell you something about the Crusades.*

The point sentence in the conclusion appears to be the last sentence:

> Thus the Crusades were not just a religious effort to recapture the Holy Land and to save God's faith, as is so widely believed in popular memory, but also a political effort to unify the Church and Europe and save them from internal forces threatening to tear them apart.

This point is more specific, more substantive, and plausibly contestable. Once we see that, we also see how to revise the last sentence of the introduction. We could simply cut and paste that concluding sentence to the end of the introduction (replacing *thus* with something appropriate, of course). Or we could write a sentence that, while not revealing the full extent of the point, would at least connect the two more clearly, like this:

> The popes urged these Crusades to restore Jerusalem to Christendom, but the documents recording their words suggest other issues as well, issues involving political concerns about European and Christian unity.

13.2.2 Step 2: Identify Your Major Sections and Their Points

The next thing that your readers must know about your organization is where one section of your paper stops and the next begins and what is the main point in each section. So in each section, do what you just did for the whole paper.

1. Divide the body of your paper into its major sections. Put an extra space between sections. If you cannot find section boundaries, your readers won't either.
2. Put a slash mark after the introduction to each major section. Each section must have a short segment that introduces it.
3. Put a slash mark before the conclusion to each major section. If your sections are shorter than a couple of pages, they may not need separate conclusions.
4. Locate and highlight the major point of each section, the sentence that expresses its main idea. If you cannot find a sentence that expresses your point, neither will your readers.
5. Ordinarily the point sentence of each section should be the last sentence of a brief introduction to that section. If the point sentence for each section is not in the introduction to that section, you must have a good reason for putting it at the end. When readers do not see the point of a section early, they have to work harder to assemble your argument.
6. *Never* locate the only point of a section in the middle of the section.

If you cannot perform each of these steps quickly, you have probably uncovered a problem with the organization of your paper. Look again at pp. 104–6, 143–44 to review how you arranged your ideas and structured your argument.

When you highlighted your points, you produced an outline

that you can now read off the page, but it would be better to write it out. Your outline will now be a list of sentences that looks like this:

> Point sentence at the end of the introduction
>> Sub-point sentence 1
>>> sub-sub point sentence 1
>>> sub-sub point sentence 2
>>> sub-sub point sentence 3
>> Sub-point sentence 2
>>> sub-sub point sentence 1
>>> sub-sub point sentence 2
>> Sub-point sentence 3
>>> sub-sub point sentence 1
>>> sub-sub point sentence 2
>> Sub-point sentence X . . .
> Main-point sentence in the conclusion

Now ask yourself, if I assembled all of these sentences into a single paragraph, would it make sense?

13.2.3 Step 3: Diagnose the Continuity of Your Themes

Your next step is to determine whether these points and sub-points "hang together" conceptually. You first have to determine whether your points are stitched together by a handful of key thematic concepts. Those are the words that express central concepts that should run from your introduction through the body into your conclusion. Conduct this test as follows:

1. In your introduction and conclusion, particularly in their point sentences, circle the key concepts that you will develop. Ignore general words like "topic," "issue," "important," "significant," and any other words that do not refer to the substance of your claim.

2. If you cannot find any key words, or only a few,

   • Look closely at the last few pages of your report for the concepts that appear there most often.
   • Incorporate those concepts into your two point sentences, both at the end of your introduction and in your conclusion.

For example, when we checked for key thematic concepts in the paper about the Crusades, we found that the point sentence in the introduction was empty of significant concepts:

I will discuss the reasons that these popes gave for a Crusade.

However, we can find several key terms in the last paragraph (and in several before):

As we can see from these documents, Popes Urban II and Gregory VII urged the Crusades as a way not just to restore the Holy Land to Christian rule, but also to **preserve** the **political unity** of the Church and Western Europe. Urban wanted to conquer the Muslims, but no less importantly to **reinforce his authority and to control fighting** among Europeans by directing their energies elsewhere. Gregory wished to **unify** the Roman and Greek Churches, but also to **prevent the breakup** of the Catholic Church and even the Empire. To achieve their **political** ends, each pope tried to **unite** people in a **common religious fight** against the East to prevent them from **fighting among themselves** and to **unify an increasingly divided** Church. Thus the Crusades were not just a **religious effort** to recapture the Holy Land and to save God's faith, as is so widely believed in popular memory, but also a **political effort to unify** the Church and Europe and save them from **internal forces** threatening to **tear them apart.**

We can assemble the key concepts under just a few terms:

preserving internal political unity, by **redirecting internal turmoil** toward an **external religious effort.**

In your introduction, readers must recognize the central concepts that you will use to stitch your paper together, and after they finish, they must leave your conclusion with those concepts etched in their memory. If the key words in your introduction are not clearly related to the key words in your conclusion, readers may feel that you have broken the implied promise your introduction made. If the circled terms in your conclusion are more detailed than those in your introduction, look closely to determine whether you should have introduced those terms in your introduction.

The next step is to trace whether those circled key terms appear consistently in the sub-point sentences in the rest of your outline. We do not have the space to illustrate the next steps here, but you should do for each section exactly the same thing that we just did with the introduction and conclusion to the paper about the Crusades:

1. Circle words in the sub-point sentences that are the same or obviously related to the circled terms in the point sentences in your introduction and conclusion.

2. If any sub-point sentence does not have terms from the introduction or conclusion, you may have failed to relate that sub-point to your main claim. Even if you think you did, your readers may not see the connection.

   • Try to revise the sub-points so that they include a few circled terms. If you can't, consider revising or even eliminating that section of your paper.

3. Now do the opposite. Check for important concepts in your sub-points that you did not mention in the introductory or concluding point sentences.

   • Revise to add these key terms to those point sentences.

Now create headings for each major section:

1. In the point sentence of each section, identify those key terms that appear only or most often in that section. These terms include names of key concepts or of major persons, places, and things.

2. Assemble those key terms into a phrase that uniquely identifies the section, and make that phrase the heading for that section. Do this even if, in your kind of writing, experienced researchers do not use headings. You can always delete them just before you print out your last draft. If you have the time, repeat this process for each major sub-sub-section.

13.2.4 Step 4: Diagnose the Whole

If your points "hang together" conceptually, now determine whether they "add up" to a whole that supports your main point, the main claim in your argument.

1. Read all those point sentences as if they were a single paragraph.
2. We can't give you a surefire way to know whether they add up to a whole, so this is a good time to ask a friend, relative, or roommate to listen to you give an oral presentation of your paper. Use your outline of points as a guide. Explain to your listener (or lacking an audience, yourself) the principle of your organization: is it chronological, and if so why? is it from most important to least important, and if so why?

13.3 REVISING YOUR ARGUMENT

Once you have determined that your organization is at least plausible, your next question is whether that organization expresses an argument or only a patchwork of quotations and data.

13.3.1 Identify Your Argument

1. Return to that outline of main and sub-points you assembled when you were diagnosing and revising your organization.
2. Determine whether those points are also the *major claims* that the rest of their particular sections support.

   • If they are not, you have a disjunction between the organizing *points* of your paper and the structure of *claims* in your argument.

3. In each section, identify everything that counts as primary evidence—all the summaries, paraphrases, quotations, facts, figures, graphs, tables—everything you quote from a primary or secondary source.
4. Now *ignoring* all of that, skim what's left. You are looking for the expression of *your* analysis, *your* evaluation, *your* judgment.

   • If what is left is much less than what you ignored, you may not have a substantial argument, but only a pastiche of raw data or a summary of them.
   • If there's time left, return to Chapters 7–10 and do what you can to beef up your personal contribution to your paper.

13.3.2 Evaluate the Quality of Your Argument

Now you must ask some harder questions. Assuming that your readers can at least follow the organization of your argument, what might cause them to reject it? At this point, you have to evaluate your evidence, your reservations, and what is most difficult, your warrants. You might review Chapters 7–10.

1. Is your evidence reliable and clearly connected to your claims? If you are close to your final draft, it may be too late to make your evidence more representative or precise, and if you are using all the evidence you have, its sufficiency and appropriateness may be closed questions. But you can check other features:

   • Check your data and quotations against your notes.
   • Make sure that your readers can see how each quotation and each set of data relate to your claim.
   • Check that you haven't skipped intermediate steps in an argument. (Review especially pp. 118–19.)

2. Have you appropriately qualified your argument?

   • Don't hesitate to drop into appropriate places a few well-placed hedges like *probably, most, often, may,* etc. (Review pp. 140–41.)

3. Does your text seem less like a contest between competing intellects and more like a conversation with someone interested in what you have to say, but with a mind of her own?

   • Readers want to hear reasons, not to challenge you but because they want to know more. *Why do you believe that? But what about . . . ? Are you really making that strong a claim? Could you explain how this evidence relates to your claim?* Go through your argument, asking such questions in unexpected places. (Review p. 144.)

4. The hardest question: What warrants have you left unexpressed?

   • Even if your readers accept your evidence as reliable, what else must they believe before they accept your claims? (Review pp. 111–26.) There is no easy way to test this. Once you have identified each section and

sub-section of your argument, write in the margin the
most important unstated warrant that you think readers
must accept. Then ask whether they will, or whether
you have to argue for it explicitly.

13.4 THE LAST STEP

In the Quick Tip on speedy reading after Chapter 6, we described a way to skim your sources for their gist so that you could
decide whether they offered anything you might find useful. Give
someone else your paper to skim in the same way, then report its
gist to you. If that reader can skim your paper easily and report its
gist accurately, you probably have a well-organized paper. If not, . . .

TITLES

The first thing your readers read—and probably the last thing you should try to write—is your title. Most writers just attach a few words that suggest what a report is "about." But a title can be more useful if it creates the right expectations, deadly if it doesn't. Here are three titles for a paper about school desegregation. Which creates the most specific expectations?

The "Separate-but-Equal" Doctrine

Economic Effects of the "Separate-but-Equal-Doctrine"

Equal Rights, Unequal Education:
Economic Racism as a Source of the "Separate-but-Equal"
Doctrine

Your title should introduce your key concepts. If, like the first title, yours merely announces a general topic, you give your readers little direction about where you are taking them. The last title announces key terms that readers can watch for. When they see them return, they will feel that the text has met their expectations.

When you must create a title, do this:

- Locate your main point sentences, either at the end of your introduction or in your conclusion (or both).
- In those point sentences, circle the words that name the most important and specific conceptual themes, abstract concepts, value judgments, etc.
- Underline the main persons, places, and things.
- Out of those two sets of words, create a two-part title that suggests your challenging question or its challenging answer. This gives you two shots at helping your readers: if you don't get it right in the first line, you might in the second. Of course, if you find the perfect one-part title, use it.

If your point sentence is vague, you are likely to end up with a vague title. If so, you will have failed twice: You will have offered readers both a useless title and useless point sentences. But you will also have discovered something more important: your paper needs more work.

ABSTRACTS

In some fields, particularly in the natural and social sciences, your report should begin with an abstract, a brief summary that tells readers what to expect. Although it should be shorter than an introduction, a useful abstract shares three key features of an introducton:

- It states the research problem.
- It announces key themes.
- It ends with a statement of the main point or with a launching point that anticipates the main point in the full text.

As with other aspects of reports, abstracts differ in different fields. But most follow one of three patterns. You can find out which patterns are used in your field by asking your teacher or looking in a standard journal.

Context + Problem + Main Point

This kind of abstract is an abbreviated introduction, beginning with a sentence or two to establish the context of previous research, a sentence or two to state the problem, and then the main result of the rescarch.

> Computer folklore has long held that character-based user interfaces promote more serious work than do graphical user interfaces (GUI), a belief that seemed to be confirmed by Hailo (1990). But Hailo's study was biased by the same folklore that it purported to confirm. In this study, no significant differences were found in the learning or performance of students working with a character-based interface (MS DOS) and students working with a graphical interface (Macintosh OS).

Context + Problem + Launching Point

This pattern is the same as the previous one, except that the abstract states not the specific results achieved, only their general nature:

Computer folklore has long held that character-based user interfaces promote more serious work than do graphical user interfaces (GUI), a belief that seemed to be confirmed by Hailo (1990). But Hailo's study was biased by the same folkore that it purported to confirm. This study tested the performance of 38 business communication students using either a character-based or a graphical interface.

Summary

In this pattern, after stating the context and the problem and before reporting the result, the abstract summarizes the rest of the report, focusing either on the evidence supporting the result or on the procedures and methods used to achieve it.

Computer folklore has long held that character-based user interfaces promote more serious work than do graphical user interfaces (GUI), a belief that seemed to be confirmed by Hailo (1990). But Hailo's study was biased by the same folklore that it purported to confirm. In this study 38 students in the same technical communication class were randomly assigned to one of two computer labs, one with character-based interfaces (MS DOS) and the other with graphical interfaces (Macintosh OS). Documents produced in the class were evaluated on three criteria: content, format, and mechanics. There was no significant difference between the two groups on any of the three criteria.

Remember that in years to come, when you have published your research, some researchers will probably look for exactly the kind of research you have done. That search will almost certainly be done by a computer looking for combinations of key words in titles and abstracts. So when you write your title and your abstract, imagine yourself as someone looking for research of exactly the kind you have done. What words is a researcher likely to be looking for? Do they appear in your title and abstract?

CHAPTER FOURTEEN

Revising Style:

Telling Your Story Clearly

Until now, we have urged you to focus more on the content and organization of your report than on its sentences. But effective sentences are also essential to a good report. When you do revise for style, preferably toward the end of the process, the steps in this chapter will help you do it efficiently.

GOOD RESEARCH REPORTS TELL A STORY, which supports a point, which solves a research problem. An important step toward that end is to be certain that your readers understand the form of your paper so that they can follow your argument. But to follow your argument, they have to understand the sentences that convey it. The problem with anticipating how your readers will judge your style, though, is that you can't do that just by reading what you've written.

14.1 JUDGING STYLE

If you had to read a long report written like (1a), (1b), or (1c), which would you choose?

1a. Too precise a specification of information processing requirements incurs a risk of a decision-maker's over- or underestimation, resulting in the inefficient use of costly resources. Too little precision in specifying needed processing capacity gives no indication with respect to the means for the procurement of needed resources.

1b. A person who makes decisions often specifies what he needs to process information. He may do so too precisely. He may overestimate the resources that he needs. When he does that, he may use costly resources inefficiently. He may also be not precise enough. He may not indicate how others should procure those resources.

1c. When a decision-maker specifies too precisely the resources he needs to process information, he may over- or underestimate them and thereby use costly resources inefficiently.

But if he is not precise enough, he may not indicate how
those resources should be procured.

Very few readers choose (1a), some choose (1b), most choose
(1c). Version (1a) sounds like a machine speaking to a machine (it
actually appeared in a respectable journal). (1b) is clearer but almost
simpleminded, like a patient adult speaking slowly to a child. (1c)
is clearer than (1a), but not condescending like (1b); it sounds like
a colleague speaking to a colleague.

We believe that researchers should usually take as their model
the style of (1c). Some disagree, claiming that heavy thinking demands heavy writing, that some ideas are so intrinsically complex
that when writers try to make them clear, they oversimplify, sacri-
ficing nuances and complexity of thought. If readers can't understand, well, that's their problem.

Perhaps. But we believe that such complex thinking appears in
print less often than most researchers think, that complex sentences
are more likely to indicate thinking that is not complex but sloppy,
and that even when thinking is so complex that it requires a complex style, those sentences always profit from a careful look.

Of course, different problems with style afflict writers at different stages in their careers. High school seniors more often write in
the simplistic style of (1b). Advanced students have problems that
develop only when they begin to specialize in a particular field, and
when that happens they often fall into a style that is almost a
parody of (1a).

In what follows, we concentrate on those matters of style that
afflict writers who are not quite beginners. We assume that you do
not need help with spelling and subject-verb agreement, so we will
not address basic grammar and usage nor a style that is too simple.
If you have problems with those matters, you'll need other assistance. We address here the problem of a style that is too complex,
too "academic," the kind of writing that typically afflicts not only
those just getting into serious research, but too many experienced
researchers as well.

This problem emerges among somewhat advanced students for
two reasons. First, when writers begin wrestling with ideas that
test their comprehension, their style breaks down in predictable

ways. Second, it is also at this point that they are starting to read journal articles and monographs written in a style so wretched that it tests the patience of even the most experienced reader. Many beginners imitate that style, thinking that it must stand for academic success. (They are wrong.)

Thus it happens that those starting advanced work are hit by double trouble. Their own prose suffers because they do not fully understand what they are reading, and the style of what they are reading is partly responsible for that suffering, but they imitate it nonetheless.

14.2 A First Principle: Stories and Grammar

When you distinguished among the styles of the three examples above, you probably used words like *clear* and *unclear, concise,* and *wordy, direct* and *indirect.* Here is an important point about those judgments: those words do not refer to the sentences that you saw on the page but to how you felt about them, to your *impressions* of them. If you said that (1a) was wordy, you were really saying that *you* had to read too many words for too little meaning; if you said (1c) was clear, you were saying that *you* found it easy to understand.

There is nothing wrong with impressionistic language, but it does not explain *what on the page makes you feel as you do.* To do that, you need a way of talking about sentence style that lets you connect your impressions to what causes you to have them.

The principles that distinguish the felt complexity of (1a) from the mature clarity of (1c) are few and simple. Those principles will direct your attention to only two parts of your sentences: to the first six or seven words and to the last four or five. If you can get those few words straight, the rest of the sentence will usually take care of itself. To understand these principles, though,

> Note that here we're talking about "revision." In Chapter 11, we urged you to draft quickly, concentrating on getting something on paper, not correcting details of sentence structure, punctuation, or spelling. If you try to apply our advice about revision *as you draft,* you will tie yourself in knots. Save your concern for style until you have something to revise.

you must first understand five grammatical terms: *subject, verb, noun, preposition,* and *clause.* (If you have not thought about those terms for a while, you might refresh your memory before you go on.)

14.2.1 Subjects and Characters

The first principle may remind you of something you learned in the ninth grade, but it is in fact more complicated. At the heart of every sentence are its subject and verb. At the heart of every story are its characters and actions. In the ninth grade, you probably learned that subjects are characters (called "doers"). But that is not always true, because subjects can refer to things other than characters. Compare these two sentences (the whole subject in each clause is underlined):

2a. <u>Locke</u> frequently repeated himself because <u>he</u> did not trust words to name things accurately.

2b. <u>The reason for Locke's frequent repetition</u> lies in his distrust of the accuracy of the naming power of words.

The subjects in (2a) fit that ninth-grade definition: the subjects— *Locke* and *he*—are doers. On the other hand, the subject of (2b)— *The reason for Locke's frequent repetition*—certainly does not, because it is not a character.

We can see the same difference between these two:

3a. If <u>rain forests</u> are continuously stripped to serve short-term economic gain, <u>the entire biosphere</u> may be damaged.

3b. <u>The continuous stripping of the rain forest in the service of short-term economic gain</u> could result in damage to the entire biosphere.

In the clearer version, (3a), look at the first few words of each clause:

3a. If <u>rain forests</u>$_{subject}$ are continuously stripped$_{verb}$ · · · <u>the entire biosphere</u>$_{subject}$ may be damaged$_{verb.}$

Their subjects name main characters: *rain forests* and *biosphere.*

3b. <u>The continuous stripping of the rain forest in the service of short-term economic gain</u>$_{subject}$ could result$_{verb}$ in damage to the entire biosphere.

In (3b), the subject does not express a character but rather an action: *The continuous **stripping** of rain forest in the service of short-term economic gain.*

If we can also agree that (2a) and (3a) are clearer than (2b) and (3b), then we can see why those ninth-grade definitions may be poor language theory but are good advice about writing clearly. The first principle of clear writing is this:

> Readers will judge your sentences to be clear and readable to the degree that you can make the subjects of your verbs name the main characters in your story.

14.2.2 Verbs, Actions, and "Nominalizations"

A second key difference between prose that seems clear and prose that seems difficult is how writers express the crucial actions in their stories—as verbs or as nouns. For example, look again at the pairs of sentences (2) and (3). (We boldface words that name actions; if those actions are verbs, we also underline them; if those actions are nouns, we double-underline them.)

2a. Locke frequently **repeated** himself because he did not **trust** words to **name** things accurately.

2b. The reason for Locke's frequent **repetition** lies in his **distrust** of the accuracy of the **naming** power of words.

3a. If rain forests are continuously **stripped** to **serve** short-term economic gain, the entire biosphere may be **damaged**.

3b. The continuous **stripping** of the rain forest in the **service** of short-term economic gain could result in **damage** to the entire biosphere.

Why are sentences (2a) and (3a) clearer than (2b) and (3b)? Partly because their subjects are characters but also because all of their crucial actions are expressed, not as nouns, but as verbs—*repeated* vs. *repetition,* the verb *trust* vs. the noun *trust;* the verb *name* vs. *naming power, strip* vs. *stripping, serve* vs. *service,* the verb *damage* vs. the noun *damage.*

Another example. This time look at the prepositions in (4a) that do not appear in (4b):

4a. Our development and standardization **of** an index **for** the measurement **of** thought disorders has made possible quantification **of** response as a function **of** treatment differences.

4b. Now that we have developed and standardized an index to measure thought disorders, we can quantify how patients respond to different treatments.

Those five prepositions—four *of*'s and a *for*—resulted directly from turning the verbs *develop, standardize, measure, quantify, respond* into the nouns *development, standardization, measurement, quantification, response.*

There is a technical term for what we do when we turn a verb (or an adjective) into a noun: we *nominalize* it. When we nominalize the verb *nominalize* we create the nominalization, *nominalization.* Most nominalizations end with syllables like *-tion, -ness, -ment, -ence, -ity.* But some of these nouns are spelled just like the verb. Some examples:

Verb	Nominalization	Adjective	Nominalization
decide	decision	precise	precision
fail	failure	frequent	frequency
resist	resistance	intelligent	intelligence
delay	delay	specific	specificity

When you nominalize adjectives and verbs in a sentence, you change that sentence in three other ways:

- You have to add prepositions.
- You have to add verbs, which will always be less specific than the ones you could have used.
- You are likely to make the characters in your story modifiers of nouns or drop them from a sentence altogether.

When we used the nominalizations in (4a) instead of the verbs in (4b), we had to add an empty verb *made,* and we squeezed *we* into *our* and dropped out *patients* altogether. And as a consequence, we created a wordier, less clear sentence.

So here are two fundamental principles of a clear style:

(1) Make your central characters the subjects of your verbs,

(2) Use verbs to express crucial actions.

14.2.3 Diagnosis and Revision

From these principles of reading, we can offer two principles of writing, one for diagnosis and one for revision:

To diagnose:

1. Draw a line under the first six or seven words of every clause, whether that clause is main or subordinate; whether at the beginning, middle, or end of a sentence.

2. If in those first six or seven words the subjects consistently refer not to characters but to abstractions or if the verb does not name a clear action, then that sentence is one which your readers may wish you had revised.

To revise:

1. First, locate in the sentence the characters you want to tell a story about. If you can't find any, decide who ought to be main characters.

2. Next, look for what those characters are doing. If their action is in a nominalization, change it into a verb (i.e., "de-nominalize" it) and make the character its subject.

3. You may have to recast your sentence around some version of *If X, then Y; Because X. . . . , Y; Although X, Y; When X, Y.*

That's the simple version. Now we make it a bit more complex.

14.2.4 Who or What Can Be a Character?

You may have been surprised when we called *the rain forest* and *the biosphere* "characters," because we usually think of characters as flesh-and-blood. In fact, most readers do prefer to read prose in which characters are flesh-and-blood people.

But we can also tell stories whose characteristics are abstractions. In your kind of research you may have to tell a story about *demographic changes, social mobility, unemployment,* or *isotherms, magnetism,* and *gene pools.* Sometimes, you have a choice: your paper in economics might tell a story about people, such as *consumer, the Federal Reserve Board,* and *Congress,* or about abstractions associated with them such as *savings, fiscal policy,* and *legislation.*

5a. When consumers save more, the Federal Reserve adopts a fiscal policy that influences how Congress legislates on taxes.

5b. Increased savings result in a Federal Reserve policy that influences Congressional tax legislation.

In this sense, a character is any entity, real or abstract, that you focus on through several sentences. A passage might be about people or about the abstractions we associate with them: *bankers* vs. *fiscal policy, savers* vs. *micro-economics,* or *analysts* vs. *predictions.* In the abstract stories that experts like to tell, main characters are often abstract nominalizations:

6. Now that we have developed and standardized an index to measure thought disorders, we can quantify how patients respond to different treatments. **These measurements** indicate that **treatments** requiring long-term **hospitalization** do not effectively reduce the number of psychotic episodes among schizophrenic patients.

The nominalizations in that second sentence—*measurement, treatment, hospitalization*—refer to three concepts as familiar to its intended readers as *doctors* and *patients.* Given that audience, the writer would not need to revise the second sentence.

That may seem to undercut our principle about getting rid of nominalizations. In a way it does, because now instead of revising every nominalization, we have to choose which to change into verbs and which to leave alone. For example, the nominalizations in the second sentence of (6) are the same as those in (7a):

7a. The **hospitalization** of patients without appropriate **treatment** results in the unreliable **measurement** of outcomes.

But that sentence would profit from revision:

7b. When we **hospitalize** patients but do not **treat** them appropriately, we cannot **measure** outcomes reliably.

So what we offer here is no iron rule of writing, but rather a principle of diagnosis and revision that you must apply judiciously.

14.2.5 Abstractions and Characters

The real problems of abstract prose occur when you create a main character out of a nominalization, use that nominalized character in the subjects of your sentences, but then sprinkle around them still more nominalizations. Here is a passage about two abstract characters, "immediate intention" and "prospective intention." Those characters are puzzling enough, but note all the other nominalizations in the same passage, complicating that story even more (We underline subjects, boldface nominalizations other than "intention"):

8a. The **argument** is this. <u>The cognitive component of intention</u> exhibits a high degree of **complexity**. <u>Intention</u> is temporally divisible into two: prospective intention and immediate intention. <u>The cognitive function of prospective intention</u> is the **representation** of a subject's similar past **actions**, his current situation, and his course of future **actions**. That is, <u>the cognitive component of prospective intention</u> is a **plan**. <u>The cognitive function of immediate intention</u> is the **monitoring** and **guidance** of ongoing bodily **movement**. Taken together these <u>cognitive mechanisms</u> are highly complex. <u>The folk psychological notion of **belief**</u>, however, is an attitude that permits limited **complexity** of content. Thus <u>the cognitive component of intention</u> is something other than folk psychological **belief**.

We can revise this to keep the abstract character "intention," but if we change unnecessary nominalizations back into verbs and adjectives (they are boldfaced), we create a much clearer passage:

8b. <u>My argument</u> is this. <u>The cognitive component of intention</u> is quite **complex**. <u>Intention</u> is temporally divisible into two kinds: prospective intention and immediate intention. <u>The cognitive function of prospective intention</u> **is to represent** how <u>a person</u> has **acted** similarly in the past, her current situation, and how <u>she</u> will **act** in the future. That is, <u>the cognitive component of prospective intention</u> lets her **plan** ahead. <u>The cognitive function of immediate intention</u>, on the other hand, lets her **monitor** and **guide** her body as <u>she</u> **moves** it. Taken together <u>these cognitive mechanisms</u> are too **complex** to explain in terms of what <u>folk psychology</u> would have us **believe**.

The point: Don't avoid nominalizations just because they are nominalizations. Some of your central characters may have to be abstractions. But if they are, avoid other nominalizations that you do not need. As always, the trick is knowing which you need and which you don't—just remember that you usually need fewer than you think. Judging how many is a skill that comes only from practice and experience.

14.2.6 Picking Main Characters

Having qualified our principle once, we complicate it a last time. If your sentences are readable, your characters will be the subjects of verbs and those verbs will express the crucial actions those characters are involved in. But most stories have several characters, any one of whom we can make more important than others simply by the way we construct sentences. Take our sentence about the rain forest:

9. If <u>the rain forests</u> continue to be stripped to serve short-term economic interests, <u>the entire biosphere</u> may be damaged.

This sentence tells a story that implies other characters but does not specify them: who is stripping the forests? More important, does it matter? This story could focus on them, but who are they?

9a. If <u>developers</u> continue to strip the rain forests to serve short-term economic interests, <u>they</u> may damage the entire biosphere.

9b. If <u>loggers</u> continue to strip the rain forests to serve short-term economic interests, <u>they</u> may damage the entire biosphere.

9c. If <u>Brazil</u> continues to strip the rain forest to serve short-term economic interests, <u>it</u> may damage the entire biosphere.

Which is best? It depends on whom the story *should* be about. As you diagnose sentences, you have two decisions. Whenever possible, put characters in subjects and actions in verbs. But be sure that the character is your *central* character, if only for that sentence.

14.3 A SECOND PRINCIPLE: OLD BEFORE NEW

There is a second principle of reading, diagnosis, and revision even more important than the one we have just explored. Fortunately, the two principles are related. Compare the (a) and (b) versions in these two pairs. Which seems easier to get through? Why? (Hint: Look at the way the sentences begin.)

> **How Necessary Is Abstraction?**
>
> If you are doing advanced work for the first time, you may think you have to write in a difficult style to sound like an expert. It's an understandable impulse. But in every field, readers prefer prose that is readable without being Dick-and-Jane simple. Your teacher wants your prose to be mature but not inflated, complex but not enveloped in smog. Some claim that they must write in a complex style to be published. We can only point to the best journals, all of which publish clearly written articles (regrettably, along with many that are not). When both are published, why choose to make your writing less readable?

10a. Because the naming power of words was distrusted by Locke, he repeated himself often. Seventeenth-century theories of language, especially Wilkins' scheme for a universal language involving the creation of countless symbols for countless meanings, had centered on this naming power. A new era in the study of language that focused on the ambiguous relationship between sense and reference begins with Locke's distrust.

10b. Locke often repeated himself because he distrusted the naming power of words. This naming power had been central to seventeenth-century theories of language, especially Wilkins' scheme for a universal language involving the creation of countless symbols for countless meanings. Locke's distrust begins a new era in the study of language, one that focused on the ambiguous relationship between sense and reference.

11a. The biosphere could be permanently damaged if rain forests continue to be stripped to serve these short-term interests. National policies that deal with local problems

and ignore the global impact will not halt this damage.
Only the efforts of the industrialized countries of the
world will achieve that goal.

11b. If rain forests continue to be stripped to serve these
short-term interests, the biosphere could be permanently
damaged. This damage will not be halted by national policies that deal with local problems and ignore the global
impact. That goal will be achieved only by the efforts of
the industrialized countries of the world.

Most readers prefer (10b) and (11b). They don't say that (10a) and
(11a) are too "complex" or "inflated," but that they seem "disjointed," they do not "flow," words that again describe not what
is on the page, but how readers *feel* about what they are reading.

We can explain what causes those impressions if we again
apply the "the-first-six-or-seven-words" test. In the "disjointed"
(a) versions, the ones that do not "flow," the sentences begin quite
differently from the sentences in the (b) versions. The sentences in
(10a) and (11a) begin with information that a reader would find
unfamiliar:

the naming power of words,
seventeenth-century theories of language,
a new era in the study of language;
national policies that deal with local problems,
an effort that involves the industrialized countries.

In contrast, the sentences in the (b) versions begin with information
that readers would find familiar:

Locke,
this naming power,
Locke's distrust (*a nominalization, but a useful one because it repeats
something from the previous sentence*);

this damage (*another nominalization that usefully repeats something
from the previous sentence*),
that goal.

These are almost all abstractions, but they refer to ideas that readers
would recall from the previous sentences.

As your readers move from one sentence to the next, they
follow your story most easily if they can begin each sentence with

a character or idea that is familiar to them, either because you have already mentioned it or because they expect it. From this principle of reading, we can infer principles of diagnosis and revision:

- Look at the first six or seven words of every sentence.
- Be certain that each opens with information that your readers will find familiar, easy to understand (usually words used before).
- Put close to the ends of your sentences information that your readers will find new, complex, harder to understand.

This principle cooperates with the one about characters and subjects, because older information usually names a character (after you have introduced it). But should it ever come to a choice between the two, *always choose the principle of old-before-new.*

14.4 CHOOSING BETWEEN ACTIVE AND PASSIVE

At this point, some of you may recall advice you once received to avoid passive verbs. That advice is not just misleading; in the sciences, it is terrible. Rather than worry about active and passive, ask a simpler question: Do your sentences begin with familiar information, preferably a main character? If you put familiar characters in your subjects, you will use the active and passive properly. For example, you may have noticed that one of our earlier examples had passive verbs:

12a. If the rain forests continue to be stripped to serve short-term economic interests, the entire biosphere may be damaged.

Had we followed the standard advice, that sentence would read:

12b. If loggers continue to strip the rain forests to serve short-term economic interests, they may damage the entire biosphere.

That sentence makes the loggers the main character—fine in a paper about logging. But if you were telling a story about the gene pool in the Amazon, then the main characters *ought* to be the rain forests and the biosphere—and so that sentence *should* be passive.

In English classes, students often hear that they should always use active verbs, but in the sciences, engineering, and some social sciences they hear the opposite—use the passive. Most of that advice (based on the alleged interest of scientific objectivity) is equally misleading.

Compare the passive (13a) with the active (13b):

13a. The fluctuations in the current <u>were measured</u> at two-second intervals.

13b. We <u>measured</u> the fluctuations in the current at two-second intervals.

These sentences are equally objective, but their *stories* differ; one is about fluctuations, the other about the person measuring. The first is supposed to be more "scientific" because it focuses on the current and ignores the person. But the passive in itself is not more objective than the active; it merely implies that the action can be performed by anonymous others who, if they want to, can replicate the researcher's procedures. So in this case, the passive is the right choice.

On the other hand, consider this pair of sentences:

14a. It is suggested that the fluctuations resulted from the Burnes effect.

14b. We suggest that the fluctuations resulted from the Burnes effect.

In this context, the active *we* is not only common in scientific prose, but appropriate. The difference? It has to do with the kind of action the verb names. The passive is appropriate when authors refer to actions they perform in the laboratory and that they encourage others to replicate: *measure, record, combine,* and so on. But when authors refer to actions that only *they* are entitled to perform—rhetorical actions such as *suggest, prove, claim, argue, show,* and so on—then the authors are the main characters and so they *should* be the subjects of active verbs. Researchers typically use the first person at the beginning of journal articles, where they describe how *they* discovered their problem, and at the end where they describe *their* solution to it.

14.5 A LAST PRINCIPLE: COMPLEXITY LAST

We have concentrated on how clauses begin. Now we'll look at how they end. You can anticipate the principle: If old information goes first, the newest, most complex information goes last. This principle is particularly important in three contexts:

- when you introduce a new technical term;
- when you present a unit of information that is long and complex;
- when you introduce a concept that you intend to develop in what follows.

14.5.1 Introducing Technical Terms

When you introduce a technical term that your readers might be unfamiliar with, construct your sentence so that your technical term appears in the last words. Compare these two:

15a. Calcium blocker drugs can control muscle spasms. Sarcomeres are the small units of muscle fibers in which these drugs work. Two filaments, one thick and one thin, are in each sarcomere. The proteins actin and myosin are contained in the thin filament. When actin and myosin interact, your heart contracts.

15b. Muscle spasms can be controlled with drugs known as calcium blockers. Calcium blockers work in small units of muscle fibers called sarcomeres. Each sarcomere has two filaments, one thick and one thin. The thin filament contains two proteins, actin and myosin. When actin and myosin interact, your heart contracts.

14.5.2 Introducing Complex Information

When you express a complex bundle of ideas that you have to state in a long phrase or clause, locate that complexity at the end of its sentence, never at the beginning. Compare these two passages:

16a. There is a second reason historians have concentrated on Darwin rather than Mendel. Hundreds of letters, both personal and scientific, to scores of different recipients, including leading scientific figures, illuminate Darwin's genius. Only ten letters to the botanist Karl Nageli, and a handful to his mother, sister, brother-in-law, and nephew, represent Mendel.

16b. Historians of science have concentrated on Darwin rather than Mendel for a second reason. Darwin's genius is illuminated by hundreds of letters, both personal and scientific, to scores of different recipients, including leading scientific figures. Mendel is represented by only ten letters to the botanist Karl Nageli, and a handful to his mother, sister, brother-in-law, and nephew.

In (16a) the second and third sentences begin with complex units of information, subjects that run on for at least two lines. In contrast, the subjects in (16b) are short, simple, easy to read—because the passive verbs (*is illuminated* and *is represented*) allowed us to move the short and familiar information to the beginning and the long and complex part to the end. (That's a main purpose of the passive verb.)

If you can recognize when phrases and clauses are complex (not easy to do, because you will be too familiar with your own prose), try to put that complexity not in the beginnings of your sentences, but at their ends.

14.5.3 Introducing What Follows

When you are introducing a paragraph, or even a whole section, construct the first sentence of that paragraph so that the key terms of the paragraph are the last words of that sentence. Which of these two sentences would best introduce the excerpt that follows?

17a. The political situation changed, because disputes over succession to the throne caused some sort of palace revolt or popular revolution in seven out of eight reigns of the Romanov line after Peter the Great.

17b. The political situation changed, because after Peter the Great seven out of eight reigns of the Romanov line were plagued by turmoil over disputed succession to the throne.

The problems began in 1722, when Peter the Great passed a law of succession that terminated the principle of heredity and required the sovereign to appoint a successor. But because many Tsars, including Peter, died before they named successors, those who aspired to rule had no authority by appointment, and so their succession was often disputed by lower-level aristocrats. There was turmoil even when successors were appointed.

Context counts for much here, but of the hundreds of readers who have been shown these passages, most feel that (17b) is both more emphatic and more cohesive with the rest of the passage. The last few words of (17a) seem relatively unimportant (in a different context, of course, they might be important) and do not introduce the passage that follows as well as (17b).

Therefore, once you've checked the first six or seven words in every sentence, check the last five or six, as well. If those words are not the most important, complex, weighty, revise so that they are. Look hard at the ends of sentences that introduce paragraphs or even sections.

14.6 SPIT AND POLISH

We've focused on issues of style special to writing research reports, on principles of diagnosis and revision that help us make prose about inherently complex topics as readable as possible. There are other principles—sentence length, the right choice of words, concision, and so on. But those are issues in writing of all kinds and are addressed by many books. And, of course, readability is not enough. After you have revised style, structure, and argument, you still have to correct grammar, spelling, punctuation, and citation form. Though important, those matters do not fall within the purview of this book. You can find help in many handbooks.

Quick Tip:
The Quickest Revision

Our advice about revision may seem nitpicky, but if you revise in steps, it is not difficult to follow. The first step is the most important: as you draft, forget about these steps (except for this one). Your first job is to create something to revise. You will never do that if you keep asking yourself whether you should have just used a verb or a noun.

If you don't have time to scrutinize every sentence, start with passages where you remember having the hardest time explaining your ideas. Whenever you struggle with content, you are likely to tangle up your prose as well. With mature writers that tangle usually reflects itself in a too complex, "nominalized" style.

For Clarity
Diagnose

1. *Quickly* underline the first five or six words in every sentence. Ignore short introductory phrases such as *At first, For the most part,* etc.
2. Now run your eye down the page, looking only at the sequence of underlines to see whether they pick out a consistent set of related words. The words that begin a series of sentences need not be identical, but they should name people or concepts that your readers will see are clearly related. If not, you must revise.

Revise

1. Identify your main characters, real or conceptual. They will be the set of named concepts that appear most often in a passage. Make them the subjects of verbs.
2. Look for words ending in *-tion, -ment, -ence,* etc. If they appear at the beginnings of sentences, turn them into verbs.

For Emphasis
Diagnose

1. Underline the last three or four words in every sentence.
2. In each sentence, identify the words that communicate the

newest, the most complex, the most rhetorically emphatic information; technical-sounding words that you are using for the first time; or concepts that the next several sentences will develop.

Revise

Revise your sentences so that those words come last.

CHAPTER FIFTEEN

Introductions

This chapter discusses introductions in a way that beginning researchers may find too detailed for their needs. Intermediate and advanced researchers, however, will find that it helps them put a crucial finishing touch on their papers, dissertations, or books. We develop ideas introduced in Chapters 3 and 4.

ONCE YOU HAVE A REVISED DRAFT, your last creative task is to ensure that your introduction frames it so that your readers will understand where they think you are taking them. The standard suggestion to write introductions last is not bad advice, because you usually need a draft before you know what you *can* introduce. Another truism: *Open by "grabbing" your readers' attention with something snappy, then tell 'em what you're gonna tell 'em.* Not bad advice either, but neither is it very helpful. Grabbing attention is tricky—open in a way that seems cute, and you lose credibility. And some ways of telling readers what to expect are better than others. In fact, introductions are so important that we devote this entire chapter to them.

15.1 THE THREE ELEMENTS OF AN INTRODUCTION

Readers never begin with a blank slate, taking at face value each word, sentence, and paragraph as it comes. They read with expectations; some they bring with them, others you must create. The most important expectations you create are in the research problem you pose (see Chapter 4). In your first few sentences, you must convince your readers that you have discovered a research problem worth their consideration and that you may even have found its solution. An introduction should never leave them wondering, *Why am I reading this?*

Different research communities do things in different ways, however, and nowhere are those differences more apparent than in their introductions. These two look quite different:

> As part of its program of Continuous Quality Improvement ("CQI"), Motodyne Computers plans to redesign the user inter-

face for its Unidyne™ on-line help system. The specifications for the interface call for self-explanatory icons that will allow users to identify their function without an identifying label. Motodyne has three years' experience with its current icon set, but it has no data showing which icons are self-explanatory. With such data, it could determine which icons to retain and which to redesign. This report provides data for eleven icons, showing that five of them are not self-explanatory.

Why can't a machine be more like a man? In almost every episode of *Star Trek: The Next Generation,* the android Data wonders what makes a person a person. In the original *Star Trek,* similar questions were raised by the half-Vulcan Mr. Spock, whose status as a person was called into question by his machine-like logic and lack of emotion. In fact, Data and Spock are only the most recent "quasi-persons" who have explored the nature of humanity. The same question has been raised by and about creatures ranging from Frankenstein to Terminator II. But the real question is why characters who struggle to be persons are always white and male. As cultural interpreters, do they tacitly reinforce destructive stereotypes of what it is about a person that we must think of as "normal"? The model to which we all must aspire, at least if we want to be real people, seems in fact to be defined by Western criteria that exclude most of the people in the world.

The topics and audiences differ and so do the problems posed, but behind those differences is a shared rhetorical pattern that readers look for in all introductions. That common structure includes at least these two elements, in this predictable order:

- a statement of the research *problem,* including something we do not know or fully understand and the consequences we experience if we leave that gap in knowledge or understanding unresolved:
- a statement of the *response* to that problem, either as the gist of its solution or as a sentence or two that promises one to come.

And depending on how familiar readers are with the problem, they may also expect to see before these two elements one more:

- a sketch of a *context* of understanding that the problem challenges.

Thus the structure of a typically explicit introduction looks like this:

Context → Problem → Response

Seen in this light, those two introductions have the same structure:

> As part of its program of Continuous Quality Improve- *context*
> ment ("CQI"), Motodyne Computers plans to redesign the
> user interface for its Unidyne™ on-line help system. . . .
> Motodyne has three years' experience with its current icon
> set,
>
> but it has no data showing which icons are self- *problem*
> explanatory. With such data, it could determine which icons
> to retain and which to redesign.
>
> This report provides data *response*
> for eleven icons, showing that five of them are not self-
> explanatory.

> Why can't a machine be more like a man? In *context*
> almost every episode of *Star Trek: The Next Generation,* the
> android Data wonders what makes a person a person. In
> the original *Star Trek,* similar questions were raised by the
> half-Vulcan Mr. Spock. . . . The same question has been
> raised by and about creatures ranging from Frankenstein's
> monster to Terminator II.
>
> But the real question is why char- *problem*
> acters who struggle to be persons are always white and male.
> As cultural interpreters, do they tacitly reinforce destructive
> stereotypes of what it is about a person that we must think
> of as "normal"?
>
> The model to which we all must aspire, at least *response*
> if we want to be real people, seems in fact to be defined by
> Western criteria that exclude most of the people in the world.

Since the center of your introduction must be the statement of your problem, we start with that, then discuss context, and finally turn to your choice of responses.

15.2 State the Problem

In Chapter 4, we discussed how topics differ from problems—a topic is just a phrase that names a concept: *the transparency of Motodyne icons* or *Quasi-persons as interpreters of humanity*. In contrast, a full statement of a research *problem* has two parts:

1. The first part states some condition of incomplete knowledge or flawed understanding.
2. The second states the consequences of that flawed knowledge or understanding, as either its costs or the benefits of resolving it.

You have a research problem *if and only if* you and your readers agree that you and they do not know or understand something, but should. This ignorance or misunderstanding we called the *condition*—some gap in knowledge, unexplained conflict, or discrepancy, some failure to know or understand. You can state this condition directly or imply it in a direct or indirection question:

> **A Note about the Examples**
>
> We have abridged our examples because most introductions are long, sometimes as much as 15–20% of a whole paper. Your introductions should be longer than ours but display the same structures and perform the same functions.

> [Motodyne] has no data showing which icons are self explanatory . . . .
> But the real question is why these characters who struggle to be persons are always white and male.

But this condition of ignorance or misunderstanding creates a full research *problem* only if you can also convince your readers that your condition has *consequences,* either in the form of costs that neither you nor your readers want to tolerate, or if you can resolve it, benefits they would like to achieve.

> With such data, [Motodyne] could determine which icons to retain and which to redesign.

> As cultural interpreters, do they tacitly reinforce destructive stereotypes of what it is about a person that we think of as "normal"?

Generally speaking, you can't go wrong explicitly following this condition-cost pattern. But your decision is complicated because sometimes you don't have to state both condition and cost explicitly.

15.2.1 When Should You State Conditions Explicitly?

Occasionally, you tackle a problem so familiar that you can imply its condition just by naming the topic. Such familiar conditions are usually found in fields like mathematics and the natural sciences, in which some research problems are long-standing and widely known. Here, for example, is an abbreviated introduction to perhaps the most significant article in the history of molecular biology, the article in which Crick and Watson report their discovery of the double-helix structure of DNA:

> We wish to suggest a structure for the salt of deoxyribose nucleic acid (D.N.A.). This structure has novel features of considerable biological interest. A structure for nucleic acid has already been proposed by Pauling and Corey. They kindly made their manuscript available to us in advance of publication. Their model consists of three intertwined chains, with the phosphates near the fibre axis, and the bases on the outside. In our opinion, this structure is unsatisfactory. . . .

By saying that they will suggest a structure for DNA, Crick and Watson implied that their readers did not know it. They did not have to say it was unknown, because they knew that every reader appreciated their problem. (Note, though, that they did raise a problem to solve by mentioning Pauling and Corey's *incorrect* model.)

More often, however, your readers won't know the flaw in *their* knowledge or understanding that your research addresses unless you tell them. Few researchers tackle problems so important that everyone in their field is waiting for their answer. You are more likely to address a problem that you have found or even invented. If so, you have to convince your readers that you have raised a problem worth their time. To do that, you must be explicit about the conditions that occasioned it: the *particular* ignorance, error, puzzle, contradiction, misunderstanding, or discrepancy that you believe your readers should appreciate.

Even if you believe that your readers know your condition, it is a good idea to make it explicit anyway. Since understanding the problem is so important to the way readers understand your report, you risk much if you assume that they know more than they do. In fact, among beginning researchers, no failure is more common than that of failing to state conditions explicitly.

15.2.2 When Should You Spell Out Costs and Benefits?

If you want more than private satisfaction from your research, you have to think about sharing your problem so that it matters to others in your community. To do that, you must convince your readers that the incomplete knowledge or understanding you have discovered is significant because, left unresolved, it has costs, or benefits if it is resolved. In short, you must help your readers understand that it is in their interest to see you solve *their* problem.

Sometimes, your introduction will describe tangible costs that your research can help your readers avoid (review pp. 52–59):

> Last year, the River City Supervisors accepted the argument that River City would benefit if it added the Bayside development project to its tax base. That argument, however, was based on little or no economic analysis. If the Board votes to annex Bayside without understanding what it will add to the city costs, *the Board risks worsening River City's already poor fiscal situation.* Once the analysis includes the added burden to city schools, as well as the costs in bringing sewer and water service up to city code, the annexation proves less advantageous than the Board has assumed.

This is the kind of problem found in "applied" research—the area of ignorance (no economic analysis) has tangible consequences in the world (finances worsen).

In "basic" research you can formulate the same kind of problem if you explain the cost, not in dollars and cents, but as flawed knowledge or misunderstanding:

> Since 1972, American cities have annexed upscale neighborhoods to prop up tax bases, actions that have often brought disappointing economic benefits. But that result could have been

predicted had they done rudimentary economic analysis. The annexation movement is a case study of how political decisions at the local level fail to use available expert information. But what remains puzzling is why cities do not seek the expertise available. *If we can discover why cities fail to rely on basic economic analyses, we can perhaps better understand why their decision-making fails so often in other areas, as well.* This paper analyzes the decision-making process of three cities that annexed surrounding areas but ignored economic consequences.

15.2.3 Testing for Conditions and Costs

We have suggested, in Chapters 3 and 4, a test to determine how clearly you have articulated the costs of not solving your problem: locate the sentences that best state your condition of ignorance or misunderstanding and insert after them the question, *So what?* You have articulated your problem persuasively when you are certain that what comes before *So what?* plausibly elicits that question for your readers and that what follows convincingly answers it.

> Motodyne has no data showing which icons are self-explanatory. *So what?* With such data, it could determine which icons to retain and which to redesign.

> The real puzzle is why these characters who struggle to be persons are always white and male. *So what?* As cultural interpreters, they may reinforce destructive stereotypes of what it is about a person that we think of as "normal."

> The story of the Alamo differs not only in Mexican and U.S. versions but also in U.S. versions from different eras. We don't know why those stories are so different. *So what?* Ah, well, let me think . . .

Answering that question is not just difficult; it can be exasperating, even dismaying. If you fall in love with stories about the Alamo, you can pursue them to your heart's content, without having to justify your pursuit to anyone but yourself: *I just like knowing.*

But before others can appreciate your research, you have to "sell" them its significance. Otherwise, why should they spend time on it? If you are writing a paper for a class, your teacher is obliged to read it. No one else is. When you address a research community, you have to convince them that your problem is—or should be—their problem as well, that they will find in your solution not just something that interests them but something in their own best interest, if they only knew what you have discovered.

What interest might anyone recognize in a problem about stories of the Alamo? Well, if they go on not knowing how those stories have evolved, how the story plays differently in Mexican and U.S. history, how Hollywood turned the story into myth, they will not understand something more important—the relation between myth and history, the troubled history of relations between Mexico and the U.S., maybe even something about our identity as North Americans.

We must be candid, though: There will always be someone who will ask again, *So? I don't care about understanding the American experience, myth and history, relations with Mexico.* To such a response, you can only shrug and think to yourself, *Wrong audience.* Successful researchers know how to find and solve interesting problems and how to convince readers that they have. But a skill no less important is knowing where to look for a forum whose readers appreciate the kind of problem that you have solved.

However, if you are certain that your readers will know the consequences of your problem, then you might decide not to spell them out. Crick and Watson decided not to specify either costs or benefits, because they knew their readers were aware that, until they understood the structure of DNA, they would not understand genetics. Had Crick and Watson spelled those costs out, they might have seemed both redundant and condescending.

If you are tackling your first research project, no reasonable teacher will expect you to articulate your problem in such detail, because you probably do not yet know what other researchers think is significant. But if you can state explicitly *your own* incomplete knowledge or flawed understanding in a way that shows that *you* are committed to improving it, you take a big step toward substantial research. You take an even bigger step if you can explain why it is important to resolve that flawed understanding, if you can show that by understanding one thing better, you understand better something else much more important, *even if it is for you alone.*

15.3 CREATE A COMMON GROUND OF SHARED UNDERSTANDING

Before you articulate any of this, however, you might first open with a context that locates your problem in a relevant background. In that way, you help your readers understand how your problem fits into a bigger picture, how it relates to other research. If reporting research is like joining a conversation, you earn the right to join by acknowledging what others have said. In most reports, you do this by briefly summarizing current relevant research. (In fact, before some readers decide whether to read a report at all, they skim the first few paragraphs to see who the author thinks is worth citing.)

Students sometimes skimp on explaining this common ground, because they write their paper as if they could just pick up where a classroom conversation left off. Their introductions are so elliptical that only someone who had participated in the course could understand it:

> In view of the controversy over Hofstadter's failure to respect the differences among math, music, and art, it was not surprising that the response to *The Embodied Mind* was so stormy. What is less clear is what caused the controversy in the first place. I will argue that any account of the human mind must be interdisciplinary.

Don't write an introduction that only your teacher can understand. Imagine you are writing to another person who once took the same course but does not know what happened in your particular class.

15.4 UNSETTLE THE COMMON GROUND WITH YOUR STATEMENT OF THE PROBLEM

Common ground has yet another function, one we can illustrate with two introductions to a familiar tale:

> One sunny morning, Little Red Riding Hood was skipping happily through the forest on her way to Grandmother's house, when suddenly Hungry Wolf jumped out from behind a tree, frightening her very much.

One morning, Hungry Wolf was lurking behind a tree, waiting to frighten Little Red Riding Hood on her way to Grandmother's house.

Which feels more compelling? The first, of course, because it opens with a stable scene that Hungry Wolf disrupts:

Stable Context:

One morning, Little Red Riding Hood was skipping through the woods.

Disrupting Problem:

Condition: when Hungry Wolf jumped out from behind a tree,
Cost: frightening her [and, if they lose themselves in the story, little children, as well].

The rest of the story complicates that problem and then resolves it.

Improbable as it may seem, introductions to research articles adopt the same strategy. Many open with the stable context of a common ground—some apparently unproblematic account of research, an uncontested belief, a statement of the community's consensus on a familiar topic. The writers then disrupt that stable context with their problem: *Reader, you think you know something, but it is flawed or incomplete.*

Here is an introduction that opens without a common ground:

It has recently been found that the chemical processes that have been thinning the ozone layer are less well understood than once thought. (*So what?*) We may have labeled hydrofluorocarbons as the chief cause incorrectly.

As disturbing as this problem seems, we can heighten its rhetorical punch by locating it in an unproblematical context of prior research, not just to orient readers toward the topic, but specifically to create an apparently stable context that we can disrupt. That disruption is almost always signaled by *but, however, on the other hand,* or some other words that signal you are disrupting the stable situation that you just created. This implicitly signals to the reader the condition of your problem: the reader's incomplete or flawed understanding:

As we have investigated environmental threats, our understanding of many chemical processes, such as acid rain and the

buildup of carbon dioxide, has improved, allowing us to understand better their eventual effects on the biosphere. (*Sounds good.*)
But recently the chemical processes that have been thinning the
ozone layer have been found to be less well understood than once
thought. (*So what?*) We may have labeled hydrofluorocarbons as
the chief cause incorrectly. (*Well, what have you found?*)

Readers thus have two reasons to recognize their self-interest in
the problem: the problem itself, but also their having been unaware
of it.

We can create common ground by focusing on the history of
research:

> Few sociological concepts have fallen in and out of favor as fast as
> religion's alleged protective influence against suicide. Once sociology's most
> basic "law," the Protestant-Catholic difference in suicide has been ques-
> tioned both theoretically and empirically. **However,** some studies still
> find an effect of religion . . . .

Or on the problem itself:

> Problem formulation is recognized as a critical research operation,
> **yet** there is no description of its methods. Nor is there a theory
> of the variety of strategies available to the researcher. . . .

Or merely on some general understanding that must be corrected:

> The Crusades in the eleventh century are generally considered to
> have been motivated by religious zeal to restore the Holy Land to Chris-
> tendom. **In fact,** the motives were at least partly, if not largely,
> political.

All this may seem formulaic, and in a way, it is. But you will
quickly realize that you cannot follow this formula mindlessly.
When you master a rhetorical pattern, you have more than a formula for writing, even more than a rhetorical device for addressing
readers in a way they understand. You also have a tool for thinking.
By requiring yourself to work through a full statement of your
problem, you have to explore what your audience knows, what
they don't know, and in particular, what they should know. That
is not "fill-in-the-blank" drill.

In fact, this pattern introduces more than half the research papers written in the humanities and social sciences. They all look different, because each plays out this pattern in different ways, using different kinds of context, spelling out conditions and costs to different degrees and in different forms. But no pattern is more common. This sort of introduction appears less often in the natural sciences, because those communities work on more widely recognized problems. When scientists open with context, it is more often the statement of a known problem, like Crick and Watson's report on DNA. (What disrupts is their announcement of a solution.) As always, look at how writers introduce their problems in your field, then do likewise. The Quick Tip survey of contradictions at the end of Chapter 8 suggests several standard patterns of Context + Disruption:

> *It has often been claimed that some religious groups are "cults" because of how they differ from mainstream churches,*_{context} **but** if we look at those organizations from a historical perspective, it is not clear when a so-called "cult" becomes a "sect" or even a "religion."_{disruption}

15.5 STATE YOUR RESPONSE

So far, we have created this two-step model of an introduction:

1. STABLE CONTEXT, in the form of common ground (optional)
2. DISRUPTION, in the form of a problem, consisting of
   a. a condition of ignorance, error, etc.;
   b. the consequences of the ignorance (in the form of the cost of leaving that condition unresolved, or the benefit if you do resolve it).

Once you disrupt your readers' stable context, you must, of course, resolve it, either by explicitly stating the gist of your solution or by implicitly promising them that you will offer a solution by the end. Readers look for that response in the last few sentences of your introduction. You can state your response in one of two ways.

15.5.1 State the Gist of Your Solution

You can state the gist of your solution explicitly. That sentence will be, of course, your main point and main claim. When you announce your main point in your introduction, you create a "point-first" paper (even though that point appears as the *last* sentence of the introduction).

> As we have investigated environmental threats, our understanding of many chemical processes, such as acid rain and the buildup of carbon dioxide has improved, allowing us to understand better their eventual effects on the biosphere. (*Sounds good.*) But recently the chemical processes that have been thinning the ozone layer have been found to be less well understood than thought. (*So what?*) We may have labeled hydrofluorocarbons as the chief cause incorrectly. (*Well, what have you found?*) **We have found that the bonding of carbon . . . .**

15.5.2 Promise a Solution

Alternatively, you can put off stating your main point by saying only where your paper is headed, thereby implying that you will present your solution in your conclusion. This kind of response is a "launching-point" and implies a "point-last" paper:

> As scientists have investigated environmental threats, their understanding . . . has improved. But recently . . . less well understood. (*So what?*) We may have labeled hydrofluorocarbons as the chief cause incorrectly. (*Well, what have you found?*). **In this report, we describe a hitherto unexpected chemical bonding between . . . .**

This introduction launches readers into the body of the paper not with its point, with the gist of its solution, but with a sentence that anticipates a solution to come.

The weakest launching-point merely announces a topic:

> This study investigates the chemistry of ozone depletion.

If you have reason to put your point at the end of your paper, be sure that your launching-point does more than just announce your topic. It should suggest the conceptual outlines of your solution and announce a plan (or both).

> There are many designs for hydroelectric turbine intakes and diversion screens, but on-site evaluation is not cost-effective. A more viable alternative is computer modeling. **To evaluate**

the hydraulic efficiency of hydroelectric diversion screens, this
study will evaluate three computer models, Quattro, AVOC,
and Turboplex, to determine which is most cost effective in
reliability, speed, and ease of use.

As you read sources in your field, note where they tend to
state their main points—at the ends of their introductions, making
them "point-first," or in their conclusions, making them "point-
last." Then do what they do.

Some authors add one more component after their point, a
sentence or two explicitly announcing the *plan* of their paper:

In Part I, we describe the models; in II, we . . . ; and in III,
we . . . .

This is common in social science writing but less frequent in the
humanities, where many readers consider it ham-handed.

15.5.3 Special Problems with Point-Last Papers

Launching-point introductions are common in the humanities,
but beginning researchers should use them cautiously. First, you can lose
your readers if you are not clear where you are going and they
miss a step in your argument. You help them stay on track by
locating your main point at the end of your introduction. A bigger
risk in a point-last paper is that you lose yourself. If you draft an
introduction that promises a solution to a problem and *you* do not
yet know what the solution is (much less the full problem), you
are not drafting a paper; you are still exploring your project. *That
is a good thing to do. Just don't turn it in as a final draft.*

Some research communities implicitly require you to locate
your main point in your conclusion (despite their handbooks on
writing to the contrary). But in such fields, readers know where to
find main points, and so after reading the title and abstract, they
turn to the end. If you must locate your point in a section called
Conclusion, write that conclusion as if it were a second introduction, more compact than the first, without the literature review,
but sketching the problem again and then stating your solution.
(See the Quick Tip on First and Last Words pp. 252–54.)

Do not write a point-last paper simply because you fear that
if you state your main claim in your introduction, you will "give it

all away" and your reader will stop reading. If you have posed a
significant problem, your readers will not accept your solution simply because you announce it. They may entertain your answer as
plausible, but they will still want to see you substantiate it. In fact,
in the world at large, readers have little patience for research reports
that read like mystery stories.

15.6 Fast or Slow?

A final choice is how quickly to raise your problem. This depends on how much your readers know. In the following, the
writer begins quickly, announcing a consensus among well-informed
engineers "up to speed." In the second sentence, he briskly disrupts
that consensus:

> Fluid-film forces in squeeze-film dampers (SFD) are usually
> obtained from the Reynolds equation of classical lubrication the-
> ory. However, the increasing size of rotation machinery requires
> the inclusion of fluid inertia effects in the design of SFDs.

This next writer addresses equally technical concepts but begins
with more familiar ones, implying readers who know much less:

> A method of protecting migrating fish at hydroelectric
> power developments is diversion by screening turbine intakes . . .
> [another 110 words explaining screens]. Since the efficiency of
> screens is determined by the interaction of fish behavior and
> hydraulic flow, screen design can be evaluated by determining
> its hydraulic performance . . . [40 more words explaining hy-
> draulics]. This study resulted in a better understanding of the
> hydraulic features of this technique, which may guide future
> designs.

If you open quickly, you imply an audience of peers; if slowly,
readers who know less than you. If your readers are knowledgeable
and you open too slowly, you may sound as if *you* know too little;
too quickly, and you will seem inconsiderate of their needs.

15.7 The Whole Introduction

What we have described here may seem to overwhelm you
with too many choices, but remember: all these choices follow what

is in fact a simple "grammar." An introduction consists of only three positions:

Common Ground + Disruption + Resolution

almost always in that order. But there are choices:

- Common Ground is optional.
- Disruption usually has both Cost and Condition, but if your readers are familiar with your problem, it might have only one of them.
- Resolution *must* state either a Main Point or a Launching-Point, preferably the first.

1. COMMON GROUND:	**Opening Moves (see next Quick Tip)**
	• A general statement.
	• An event or anecdote.
	• A quotation or provocative fact.
	Context
	• Shared understanding about the current status of the problem or taken-for-granted background.
2. DISRUPTION:	**Denial:** *but, however, on the other hand,* etc.
	Statement of the Problem
	• CONDITION of ignorance, misunderstanding, etc.
	• COST/BENEFITS of leaving that condition unresolved or resolving it.
3. RESOLUTION:	**Statement of Response**
	Main Point *or* Launching-Point

Like all structural summaries, this one can feel mechanical. But when you flesh this pattern out in a real paper, readers lose sight of the form and notice only the substance, which the expected form in fact helps them to understand.

Quick Tip:
First and Last Words

Your First Few Words

Many writers find the very first sentence or two especially difficult to write. First, know what to avoid:

- Don't start with a dictionary entry: *Webster defines ethics as . . . .* If the word is important enough to define in a paper, it is too complex for a dictionary definition.
- Don't start grandly: *The most profound philosophers have for centuries wrestled with the important question of . . . .* If your subject is grand, let it speak for itself.
- Avoid *This paper will examine . . . , I will compare . . . .* Some published papers begin like that, but most readers find it banal.
- Remember not to repeat the language of your assignment. If you are struggling to start, prime the pump with a paraphrase, but when you revise, eliminate it.

Here are three choices for your first sentence or two.

Open with a Striking Fact or Quotation

Open with a fact or quotation only if its language leads naturally into the language of the rest of your introduction:

> "From the sheer sensuous beauty of a genuine Jan van Eyck there emanates a strange fascination not unlike that which we experience when permitting ourselves to be hypnotized by precious stones."
>
> Edwin Panofsky, who had a way with words, suggests here something magical in Jan van Eyck's works. Jan's images hold a fascination . . . .

Open with a Relevant Anecdote

Open with an anecdote only if its language or content connects to your topic. This paper addressed the economics of school segregation:

> This year, Tawnya Jones begins junior high in Doughton, Georgia. Though her classmates are mostly black like herself, her

school system is considered legally racially integrated. But except
for a few poor whites and Hispanic students, Tawnya's school
still resembles the segregated and economically depressed one
that her mother entered in 1952 . . . .

Open with a General Statement

Open with a general statement followed by more specific
ones until you reach your problem. This is just another version of
common ground.

> In the last decade, computers have found a host of surprising
> applications, many of which are transforming the human land-
> scape. One arena that has been transformed most quickly is the
> workplace. Even the most routine manufacturing processes now
> employ robots to do work that is too hazardous, too onerous,
> or too boring for human beings to undertake. . . .

A risky version of this is the *dawn of time* ploy, because you may
have to march through a lot of history to get to your point.

> Our fascination with machines that move under their own
> power is as old as recorded history. In ancient Greece plays were
> performed entirely by puppets driven by weights hung on twisted
> cords. Much later, European rulers were enthralled by automata
> that could write, draw, and play musical instruments. In the
> 19th century, . . . Early in this century, . . . Today, however,
> the aura of automata has vanished: industrial robots are used
> everywhere . . . .

If you open with any of these devices, be sure to use language that leads to your context, problem, and the gist of its
solution.

Your Last Few Words

Not every research paper has a section titled *Conclusion,* but
they all have a paragraph or two to wrap them up. You may be
happy to know that even a complex conclusion employs the same
elements as your introduction.

Close with Your Main Point

If you ended your introduction not with your main point but
with a launching-point, your conclusion will be your only chance

to state your main point fully. Be certain that its key terms
match those in your introduction. If you end your introduction
with your main point, restate it more fully in your conclusion.
Thus the first correspondence between introduction and conclusion is an echo—your conclusion echoing key terms from your
introduction.

Close with a New Significance or Application

A way to go beyond a flat-footed restatement of your claim
is to point out a significance in your problem that you did not
mention in your introduction. This new significance could earlier
have answered the question *So what?* but perhaps at a level more
general than you wanted to aim at then. In fact, as you formulate
a problem, find several answers to *So what?*, several costs to the
condition. Then set one aside that seems provocative enough to
use in your conclusion.

In this next conclusion, the writer introduces for the first
time an additional cost of the Supreme Court's decision on military death sentences: the military may have to change the culture
of its thinking.

> In light of recent Supreme Court decisions rejecting mandatory
> capital punishment, then, the mandatory death provision for
> treason in article 106 of the Universal Code of Military Justice
> is apparently unconstitutional and must therefore be rewritten.
> **More significantly, though, if this change does affect the course
> of military justice, it will challenge one of the most fundamen-
> tal values of the military culture, that the ultimate betrayal
> mandates the ultimate penalty.**

The writer could have used that implication in his introduction,
as a potential cost resulting from new Supreme Court decisions,
but he may have felt that such a point was too volatile to raise
so early. Take care not to make this more general significance
seem to be your main point. You can be clear about its role by
introducing it almost "by the way," as an additional implication
of your solution.

If your research is not motivated directly by a practical problem in the world, you might ask now whether its solution has

any application to one. Back in Chapter 4, we distinguished between research problems and practical problems by distinguishing
knowing from doing:

1. I am studying the way high school seniors approach
   writing essays
   2. because I am trying to discover how they choose
      topics
      3. in order to understand why they are unable to nar-
         row a topic to one they can handle in three pages
         4. *so that we can teach them how to pick topics that they
            can write about successfully.*

If your solution has an application, you can suggest it in the conclusion.

This is the second correspondence between your introduction
and conclusion. In your introduction, you "sold" your problem
by citing the costs of not solving it. In your conclusion, you can
elevate the significance of your solution by mentioning some new
and perhaps even unexpected benefit of clearer understanding
that your solution *might* have.

Close with a Call for More Research

If the significance of your solution is especially interesting,
you can call for more research:

> Data from patient records suggest that social and cultural
> factors such as gender, marital status, and age affected definitions
> of mental illness and assumptions about diagnoses. **If we are to
> understand the social values that affect the ideology of mental
> illness and the practice of psychiatry, historians must better
> understand institutional politics, medical theory, and the per-
> ceptions of the public.**

These are the third and fourth correspondences between your
introduction and conclusion. In your introduction, you may have
opened with background research before you introduced your
problem and then pointed out the incompleteness of that research. Here in the conclusion you can point to a remaining area
of ignorance, confusion, or uncertainty and then invite readers to
do more research to resolve it.

Close with a Coda

Finally, you can end with what we might call a "coda," a rhetorical gesture that adds nothing substantive to your argument but rounds it off with a graceful close. A coda can be an apt quotation, anecdote, or just a striking figure of speech, similar to or even echoing your opening quotation or anecdote—one last way that introductions and conclusions speak to each other. Just as you opened with a kind of prelude, so can you close with a coda. In short, you can structure your conclusion as the mirror image of your introduction:

Introduction	Conclusion
1. Opening quotation/fact.	5. Gist of solution.
2. Context of past research.	4. Larger significance/application.
3. Condition of ignorance.	3. What is still not known.
4. Cost of that ignorance.	2. Call for further research.
5. Gist of solution.	1. Closing quotation/fact.

Research and Ethics

EVERYTHING WE'VE SAID ABOUT RESEARCH begins with our conviction that it is a thoroughly social activity, one that links us to those whose research we use and in turn to those who will use ours. It is also an activity no longer confined to the small social world of the academy. Research is now at the center of industry, commerce, government, education, health care, warfare, even entertainment and religion. It influences every part of our society and our lives, public or private. Because research and its reporting have become a seamless part of our social fabric, in these last few pages we offer some brief reflections on an issue beyond its technique— the inescapable connection between reporting your research and the principles of ethical communication.

More than most social activities, research challenges us to define our ethical principles and then to make choices that violate or honor them. At first glance, the academic researcher must seem less tempted to sacrifice principle for gain than, say, a Wall Street researcher evaluating a stock that her firm wants to sell the public. No teacher will pay you to write a paper supporting a particular point, as some scientists are paid to testify that a product is safe. Nor is the vision of international fame likely to tempt you to compromise your principles as it apparently did the American researcher who claimed he had discovered an HIV virus that he had in fact "borrowed" from a laboratory in France.

Nevertheless, starting even with your first project, you face ethical choices. Some are the obvious "Thou shall not's" that we have discussed throughout:

- Ethical researchers do not steal by plagiarizing or claiming the results of others.
- They do not lie by misreporting sources or by inventing results.
- They do not destroy sources and data for those who follow.

Other principles of ethical research are less obvious, but implicit:

- Responsible researchers do not submit data whose accuracy they have reason to question.
- They do not conceal objections that they cannot rebut.
- They do not caricature those with opposing views or deliberately state their views in a way they would reject.
- They do not write their reports in a way that deliberately makes it difficult for readers to understand them, nor do they oversimplify that which is legitimately complex.

We can state these principles easily enough and apply them to the obvious offenders, like that biologist who marked his mice with india ink to make a genetics experiment turn out right or the student who submits as his own a paper pulled from his fraternity's file or the writer who deliberately writes turgid prose to make his thinking seem more profound.

More challenging, though, are those occasions when ethical principles take us beyond prohibitions and urge us to act affirmatively. Many philosophers have argued that the essential ethical problem is not just how to avoid violating obligations to others but rather how to join with them in a mutual project of developing what the Greeks called *ethos,* or character. When we think about ethical choices in this way, as a shared construction of *ethos,* we no longer face a simple choice between our own self-interest and the interests of another, but a challenge to find another way that is good for both.

In real situations, of course, these principles always force us to ask hard questions that your three authors would answer in different ways; but one thing that we agree on is that research offers every researcher ethical invitations that when accepted can serve the best interests of both the researcher and his readers. When you try to explain to others why the results of your research *should* change their knowledge, understanding, and beliefs because it is in their best interests to change them, you must closely examine not only your own understanding, but your own interests, as well. When you create, however briefly, such a community of shared understanding and interest, you set a standard for your work higher than any you would set for yourself alone. When you are sensible to the objections and reservations of your readers, you help yourself move closer to more reliable knowledge, better understanding, and sounder beliefs. When you conduct your research and prepare your report as a conversation among equals, all working to move toward new knowledge and better understanding, the ethical demands you place on yourself focus on the ultimate benefit of all concerned.

In this view, whatever speaks to the best interests of your readers, to their best habits of mind and heart, will be good for your own as well. When you set high ethical standards for your research, you join not just the community of those working on your narrow topic—how Hollywood changed the story of the Battle of the Alamo, say—but also the large and permanent community of everyone who has ever been curious, worked to satisfy that curiosity, and then shared that new knowledge with others.

It is this concern for the integrity of the work of the community that explains why researchers condemn plagiarism so strongly. Intentional plagiarism is theft, but of more than words. By not acknowledging a source, the plagiarist steals some of the little reward that an academic community has to offer, the enhanced respect that a researcher spends a lifetime trying to earn. The plagiarist steals from his community of classmates by making the quality of their work seem worse by comparison and then perhaps steals again by taking one of the few good grades reserved to reward those who do good work. By choosing not to learn the skills that research can teach her, the plagiarist not only compromises her own education but steals from the larger society that devotes its resources to training students to do reliable work later. Most important, plagiarism, like theft among friends, shreds the fabric of community. When intellectual thievery becomes common, the community grows suspicious, then distrustful, then cynical—*So who cares? everyone does it.* Teachers then have to worry as much about not being tricked as about teaching and learning.

From beginning to end, when research addresses the needs of readers, their knowledge, their place in a community, even if that community is fleeting or conflicted, it invites you to consider not just your topic, question, or problem, but also your obligations to your sources and to your readers. When you respect sources, preserve and acknowledge data that may run against your results, assert claims only as strongly as warranted, and acknowledge the limits of your certainty, you do so not just to avoid violating moral rules and gain credit. When you recognize the larger benefit that comes from building the kind of relationship with your readers that the best principles of research foster, then you will discover that research done in the best interests of others will also be in your own.

A Postscript for Teachers

WE WROTE THIS BOOK for those who believe—or will at
least consider—two propositions about learning and doing research:

- Students learn to do research well and report it clearly
  when they take the perspective of their readers and of
  the larger communities whose values and practices define competent research and its reporting.
- They learn to manage an important part of that complex mental and social process when they understand
  how a few key formal features of their texts influence
  the way their readers will read.

READING, RESEARCHING, AND DRAFTING:
A SELF-SUSTAINING PROCESS

These two propositions, we believe, are closely related. The
formal features that guide readers can also guide students through
the process of drafting by helping them to see how their text can
give readers what they want and need as they work to understand,
agreeing with a point here, objecting to another there, asking questions,
most of all trying to discover the significance the report has for them.

No less important, we believe that by understanding the com-
plementary processes of reading and writing, students can better
plan and conduct their research by anticipating what they have to
find and think about and eventually write about. By understanding
their own reading, they can as writers better anticipate the expectations of their readers. And by understanding what their readers will
look for in their reports, they learn to read the reports of others
more critically. The two processes, reading and writing, are mutu-
ally supporting.

THE RISKS AND LIMITATIONS OF FORMALISM

Our focus on the formal aspects of writing, however, is not
without its risks, especially with beginning researchers. Formal patterns can be too easily trivialized into meaningless activity by teachers who mistake form for substance. Like those who teach dancers
only to make their feet touch the right marks or pianists only to

hit the right keys, they think that if students merely learn and practice the motions of a complex and creative activity, they will understand its substance and significance and become competent at whatever they are practicing.

Throughout this book, we have tried to deflect mere mechanical performance by keeping students mindful of the significance of their work. We show them how the features we describe are not arbitrary empty forms to be mindlessly filled, but rather generative elements of their texts that not only influence how readers read but can help stimulate hard thinking in the writer. In fact, we believe that these patterns best help students recognize what is most important in the relationship between a researcher, her sources, her disciplinary colleagues, and her immediate readers, a crucial prerequisite to creative and original research.

These patterns, however, can still result in empty imitation if teachers fail to create a rhetorical context that requires students to understand their social role as researchers, at least in simulation. No textbook can do that; only the right kind of class experience can, something only teachers can provide. We can show students here the general patterns that most research follows. We can tell them that their readers will expect to see particular variations on those patterns, depending on their discipline, or even on their specific situation. But we cannot spell out those many variations and special circumstances.

Only teachers can construct assignments that create situations whose social dynamic gives point and purpose to research and whose key features students can recognize and understand. The less experience students have, the more social support teachers will need to provide before students can employ the formal patterns in genuinely productive ways.

On Assignment Scenarios: Creating a Ground for Curiosity

Teachers have found many ways of constructing research assignments that provide the social support students need. The most successful have these features:

1. Good assignments establish outcomes that go beyond creating a product merely to be evaluated.

They ask students to raise a question or problem that some

reader wants to see resolved, and to support that resolution with evidence that the reader judges to be reliable and relevant. Students learn little from a social dynamic whose only goal is to show a teacher that they can put the right pieces in the right places. Effective research assignments allow students to experience, or at least to imagine, a situation in which their readers need information that only they can provide.

The best assignments ask students to write for those who actually need to know or understand something better. Those readers might be a well-established community of researchers or a community of interest that their problem transiently creates. Students might do their research for a client outside of the class. A senior design class, for example, might address a problem of a local company or civic organization; a music class might write program notes; a history class might investigate the history of their university or local community. Less experienced students might write for their classmates, but they might also write for students in another class who could actually use the information that a beginning researcher could provide. They might do preliminary research for those senior design students or for students in a graduate seminar; or they might even write reports back to the students in their high schools.

Next best are assignments that simulate such situations, in which students assume that other students or a client or even other researchers have a problem that the student researcher will work to resolve. In many classes, groups of students can serve as readers whose interests and concerns beginning researchers can reasonably address.

2. Good assignments stipulate a familiar audience.

Students have trouble imagining the concerns of readers whom they have never met and whose situation they have never experienced. But even when readers are real, students have to know something about their situation in order to anticipate their concerns. Biology students with no knowledge or experience of working with a government agency will struggle to write a report that will meet the concerns of a state EPA administrator.

3. Good assignments create scenarios that are rich in contextual information.

When students write to resolve the problems of readers known and accessible to them, the assignment creates a scenario with all the wealth of reality. Students can investigate, interrogate, and analyze the situation as much as time and their ingenuity allow. As they work to understand the social dynamic that gives significance to the formal rhetorical patterns they are learning to deploy, they are likely to find the cues they need almost anywhere, often in places that teachers least expect.

When it is not practical to locate the project in a real context, the assignment must be as informative as possible. The more information you provide in writing the better. But since it is seldom possible to anticipate and write down everything students need to know about such a scenario, it is important to make analysis and discussion of it a part of the writing process. Only when students are working in a social context do they have meaningful choices to make and reasons to make them. Only when they have good reasons to make choices do those choices become rhetorically significant. And only when writers can make rhetorically significant choices will they understand that at the heart of every real writing project is the accurate anticipation of their readers' responses. When students are allowed no choices, either because the project has turned into a mechanical drill or because the project has no rhetorical setting, doing research and writing it up become merely make-work—for you as much as for them.

4. Good assignments provide interim readers.

Few professional researchers call a report finished before they have solicited and evaluated responses, something students need even more. In this book we encourage students to solicit early responses from colleagues, friends, family, even from their teachers. Getting responses is easier if opportunities are built into the assignment itself. Other students can play this role reasonably well, but not if they think that their task is just "editing"—which for them often means rearranging a sentence here and fixing a misspelling there. Those who provide interim responses must participate in the scenario as imagined readers.

5. As with any real project, good assignments give students time and a schedule of interim deadlines.

Research is messy, so it does no good to march students through it in a lock-step order: (1) select topic, (2) state thesis,

(3) write outline, (4) collect bibliography, (5) read and take notes, (6) write paper. That caricatures how research really works.

But most student researchers still need some framework, a schedule of tasks that helps them monitor their progress. They thus need time for false starts and blind alleys, for revision and reconsideration. They need deadlines for tasks well in advance of a final deadline, and interim stages for them to share their progress. The sequence of that schedule can be taken from the four Parts of this book.

RECOGNIZING AND TOLERATING THE INEVITABLE

There is another kind of support students need: honest recognition of what can reasonably be expected of them and tolerance for the entirely predictable forms of behavior that make more experienced practitioners wince. Beginners inevitably behave in awkward ways, taking suggestions and principles as inflexible rules that they apply mechanically. Taking these principles as rules, some students will work their way through a topic to a question to the card catalogue to a not very satisfactory conclusion, not because they lack imagination or creativity, but because they are acquiring a skill that to them is surpassingly strange. Such awkwardness is an inevitable stage in learning any skill that is the basis for creativity. We are not troubled when most of our beginning students produce reports that look like all the others. We have learned to defer for a while the gratification we take in their originality.

Also, we do not expect all students to report a fully articulated solution to the problem they pose. In fact, we assure them that even if they do not solve their problem, they will have succeeded in writing a valuable piece of research if they can just pose it in a way that convinces us that it is new and arguably *needs* solving. Supporting such a claim requires more research and more critical ability than merely answering a question. This kind of proposal paper is often more difficult to write than one in which a student can ask a question and answer it.

We know that on occasion students will want to use the research assignment simply to gather information on a topic, to review a field simply to gain control over it. On those occasions, we know that posing a significant problem will seem an artificial requirement.

In that situation, students might imagine that they have been asked by a supervisor or professor to survey a topic and write up a coherent, competent account of it for someone who is intelligent but does not have the time to do the research. In that context, making sense of a topic for someone else is the best way to make sense of it for themselves, weeks or months later, when they discover that they have forgotten much of the information that they took for granted while they were in the midst of their reading.

Finally, it is important to understand that different students stand in different relations to the research practices we teach. With advanced students, we do not hesitate to require them to reproduce the fine-grained details of our own disciplinary practices. But with beginners, we try to remember that, unlike advanced students, they have not made the same commitment to our community and to our underlying values. Some will make that commitment, but most will not. And so we expand our conception of what counts as successfully using and deploying the formal patterns that underlie all research, confident in our belief that by learning both to identify those patterns explicitly and to use them successfully in one setting, those students are a step closer to using them well when they later find the research community they want to join.

A Bibliographical Essay:

Our Sources and Some Suggestions

We have organized this book around the writing process, believing as we do that writing is not just the last stage of a research project but from its beginning a guide to critical thinking. This is a view commonly held in writing studies today. But we have embraced an aspect of writing that the common view has ignored, even rejected: rather than treat the standard forms of discourse and style as constraining and coercive, we believe that they are in fact creative and constructive, that they can motivate not just a critique, but the kind of thinking that encourages imagination and discovery.

In another reversal, we have shifted attention from the lone writer as the prime creative force to focus on the interaction of writer and reader and on how that interaction can help you to draft your work, to develop and test your argument, even to conduct your research. We believe that some of the most creative moments of research occur not when you are deciding what *you* want to put in our report but when we are thinking about what *your readers* must see there if they are to read it well and trust its conclusions.

We believe that it would not help and might confuse you if we kept citing the standard views and explaining how we followed or departed from them. So we did not cite any of the works that expound those views. Nor have we cited the monuments in the long tradition of rhetorical scholarship on which all of us rely.

We offer now this short essay to acknowledge the few sources we did use directly and to mark some bibliographical trails for those who might find the rhetoric of research interesting enough to explore as a research problem. Inevitably, we will have ignored a text that some will think is crucial to the field. But we do not try to cover the entire territory, or even to map out all of its prominent features. We aim only to mark a few trails that can take you as far as you wish, because the study of rhetoric now leads into every human science.

GENERAL BACKGROUND

Almost every contestable issue in rhetoric begins with Plato's *Phaedrus* and *Gorgias* (*Gorgias/Plato,* trans. Robin Waterfield, Oxford University Press, 1994) and Aristotle's *Rhetoric* (*On Rhetoric: A Theory of Civic Discourse,* trans. George Kennedy, Oxford University Press, 1991). (There are countless editions of these works; we cite only recent ones.) The best discussion of what rhetoric is *for* is Eugene Garver's *Aristotle's Rhetoric: An Art of Character* (University of Chicago Press, 1994). Following Aristotle is a long tradition of thought, including Cicero's *De Oratore,* trans J. S. Watson (Southern Illinois University Press, 1986) and *De Inventione,* trans. H. M. Hubbell (Harvard University Press, 1976), and Quintilian's *Institutiones oratoriae,* ed. James J. Murphy (Southern Illinois University Press, 1987). A study that traces the classical tradition into the modern world is Thomas M. Conley's *Rhetoric in the European Tradition* (University of Chicago Press, 1994).

The modern tradition begins with eighteenth-century rhetoricians such as George Campbell, *The Philosophy of Rhetoric,* ed. Lloyd F. Bitzer (Southern Illinois University Press, 1963, 1988). In the twentieth century, classic works include Chaim Perelman and Lucie Olbrechts-Tyteca's *The New Rhetoric: A Treatise on Argumentation,* trans. John Wilkinson and Purell Weaver (Notre Dame University Press, 1969); Kenneth Burke, *A Grammar of Motives* and *A Rhetoric of Motives* (both University of California Press, 1969); and Wayne Booth's *Modern Dogma and the Rhetoric of Assent* (Notre Dame University Press, 1974). Some would include in the contemporary tradition the work of post-structuralists such as Jacques Derrida, as found in *Margins of Philosophy,* trans. Alan Bass (University of Chicago Press, 1982).

Excerpts from across the tradition are in Patricia Bizzell and Bruce Herzberg's anthology, *The Rhetorical Tradition: Readings from Classical Times to the Present* (Bedford Books, 1990). A useful anthology of articles is *Essays on Classical Rhetoric and Modern Discourse,* ed. Robert J. Connors, Lisa S. Ede, and Andrea A. Lunsford (Southern Illinois University Press, 1984). A widely used textbook that interprets the classical tradition for today's writing student is Edward P. J. Corbett's *Classical Rhetoric for the Modern Student,* 3d edition (Oxford University Press, 1990). A survey of modern rhetoricians with a good bibliography is Sonja K. Foss, Karen A. Foss, and Robert Trapp's *Contemporary Perspectives on Rhetoric* (Waveland Press, 1985).

RESEARCHERS AND READERS

Rhetorical studies have always considered audiences, but only recently have they focused on particular social or disciplinary contexts, especially on how communities of researchers differ not only in their common knowledge and beliefs but also in the way their research sites and practices influence their discourse. A seminal inquiry into these matters is Bruno Latour's *Science in Action* (Harvard University Press, 1987). See also Greg Meyers, *Writing Biology* (University of Wisconsin Press, 1990), and Charles Bazerman, *Shaping Written Knowledge* (University of Wisconsin Press, 1988). Sophisticated studies about the rhetoric of particular fields include Donald McCloskey's *The Rhetoric of Economics* (University of Wisconsin Press, 1985), Alan G. Gross's *The Rhetoric of Science* (Harvard University Press, 1990), and Austin Sarat and Thomas R. Kearns' *The Rhetoric of Law* (University of Michigan Press, 1994).

Two useful anthologies of modern studies are *The Rhetorical Turn: Invention and Persuasion in the Conduct of Inquiry,* ed. Herbert W. Simons (University of Chicago Press, 1990), and *Textual Dynamics and the Professions,* ed. Charles Bazerman and James Paradis (University of Wisconsin Press, 1991). Some research on the role of social forces has focused on gender: see Evelyn Fox Keller, *Reflections on Gender and Science* (Yale University Press, 1985), and a collection, *Body Politics: Women and the Discourses of Science,* ed. Mary Jacobus, Evelyn Fox Keller, and Sally Shuttleworth (Routledge, 1990).

ASKING QUESTIONS, FINDING ANSWERS

The arts of inquiry begin with Aristotle's topoi (a near synonym for the term *warrants*) and Cicero's *De Inventione*. Among the most influential of the modern approaches to "invention" is Richard Young, A. L. Becker, and Kenneth Pike's *Rhetoric: Discovery and Change* (Harcourt Brace Jovanovich, 1970). (The scheme of questions outlined in Chapter 3 traces to Kenneth Pike's original work in tagmemics in the 1960s.) On the idea of "problem," see an old but still seminal book, John Dewey's *How We Think* (Heath, 1910). For a psychologist's point of view, see *The Nature of Creativity,* ed.

R. J. Sternberg (Cambridge University Press, 1988). For a conceptually based approach to using bibliographical sources, see Thomas Mann, *Library Research Models: A Guide to Classification, Cataloging, and Computers* (Oxford University Press, 1993).

ARGUMENTS

Our section on argument was inspired by Stephen Toulmin's *Uses of Argument* (Cambridge University Press, 1958), a book that has changed the way many rhetoricians think about the formal structure of argument. His views were expanded in a textbook written with Richard Rieke and Allan Janik, *An Introduction to Reasoning*, 2d edition (Macmillan, 1984). We should note that we have substantially modified Toulmin's layout of argument. A critique of Toulmin's approach with substantial bibliography is James B. Freeman's *Dialectics and the Macrostructure of Arguments* (Foris, 1991). There is a long history of studying argument in more traditional ways. Extensive references are in Frans H. van Eemeren, Rob Grootendorst, and Tjark Kruiger's *Handbook of Argumentation Theory* (Foris, 1987). A useful application of conventional logic to argument is in David Kennedy's *The Art of Reasoning* (Norton, 1988). A textbook that addresses many aspects of written arguments is Jeanne Fahnestock and Marie Secor's *A Rhetoric for Argument*, 2d ed. (McGraw-Hill, 1990). The general question of evidence in a variety of fields is addressed in *Questions of Evidence*, ed. James Chandler, Arnold I. Davidson, and Harry Harootunian (University of Chicago Press, 1994). The Quick Tip on contradictions following Chapter 8 was inspired by Murray Davis's "That's Interesting! Towards a Phenemenology of Sociology and a Sociology of Phenomenology," *Philosophy of the Social Sciences* 1 (1971):309–44.

DRAFTING AND REVISING

More about organization and style is in *Style: Toward Clarity and Grace* (University of Chicago Press, 1990) by Williams, including two chapters co-authored with Colomb. (A version limited to style but including exercises is Williams's *Style: Ten Lessons in Clarity and Grace*, 4th ed. (HarperCollins, 1993). Two quite different ways to think about style are Richard Lanham's *Style: An Anti-Textbook* (Yale University Press, 1974) and Walker Gibson's *Tough, Sweet and Stuffy: An Essay in Modern American Prose Styles* (Indiana University Press,

1966). The classic works in the visual presentation of data are Edward Tufte's *The Visual Display of Quantitative Information* (Graphics Press, 1983) and *Envisioning Information* (Graphics Press, 1990). Advanced students might look at William S. Cleveland's *Elements of Graphing Data* (Wadsworth Press, 1985) and his and Marilyn E. McGill's *Dynamic Graphics for Statistics* (Wadsworth, 1988). For the rhetoric of maps, see Mark Monmonier's *Mapping it Out: Expository Cartography for the Humanities and Social Sciences* (University of Chicago Press, 1993). An approach to introductions that also takes a structural view but offers a description usefully different from ours is in John Swales's *Genre Analysis: English in Academic and Research Settings* (Cambridge University Press, 1990).

ETHICS

Concern about the ethics of rhetoric is as old as rhetoric itself. Two major classical discussions are Plato's *Gorgias,* and Book XII of Quintilian's *Institutes.* The matter of rhetoric and ethics was revived in modern times by Burke's *A Grammar of Motives* and by Richard Weaver's *The Ethics of Rhetoric* (Henry Regnery, 1953), a book that still provokes controversy. A contemporary discussion of the more general notion of ethics in communication is Richard Johannesen's *Ethics in Human Communication,* 3d ed. (Waveland, 1990). A "postmodern" rhetoric has been found by some in Jürgen Habermas's *Moral Consciousness and Communicative Action,* trans. Christian Lenhardt and Shierry Weber Nicholsen (MIT Press, 1990) and Michael Foucault's *History of Sexuality,* trans. Robert Hurley (vol. 1, Vintage Books, 1980; vol. 2, Pantheon 1984; vol. 3, Pantheon 1986). Recently, feminist scholars have critiqued the traditional view of argument as conflict in ways similar to ours, questioning whether the standard forms of argument are too coercive and patriarchal ever to be ethical. For a short survey with bibliography on the general question of gender, language, and communication, see Sonja K. Foss, Karen A. Foss, and Robert Trapp's *Contemporary Perspectives on Rhetoric,* 2d edition (Waveland Press, 1990). See also *Contending with Words: Composition and Rhetoric in a Postmodern Age,* ed. Patricia Harkin and John Schilb (Modern Language Association of America, 1991). For a discussion of why our culture predisposes us to think about argument as conflict, see George Lakoff and Mark Johnson, *Metaphors We Live By* (University of Chicago Press, 1980).

SOURCES OF FURTHER BIBLIOGRAPHY

An annual bibliography for research in teaching writing appears in the journal *Research in the Teaching of English*. An annual bibliography of rhetoric and composition was *Longman Bibliography of Composition and Rhetoric,* ed. Erika Lindemann (Longman, 1987–), now continued by *CCCC Bibliography of Composition and Rhetoric* (Southern Illinois University Press, 1990–). Journals that publish nontechnical articles on these topics include *College Composition and Communication, College English, Journal of Advanced Composition, Philosophy and Rhetoric, Pre/Text, Quarterly Journal of Speech, Rhetorica, Rhetoric Review,* and *Rhetoric Society Quarterly.* More technical work appears in *Applied Linguistics, Discourse Processes, Text,* and *Written Communication.* Because rhetoric is now conceived so broadly, look at citations in bibliographies of current articles for other journals to monitor.

An Appendix on Finding Sources

THERE IS A VAST LITERATURE on finding information, only a small part of which we can list. We have divided this list into "General Sources" and "Special Sources," the "Special Sources" into "Humanities," "Social Sciences," and "Natural Sciences"; each of those areas into special fields; then for each field, six kinds of resources:

1. A dictionary that briefly defines concepts and sometimes offers bibliography.
2. An encyclopedia that gives more extensive overviews and usually bibliography.
3. A guide to finding resources in a field and using its methodology.
4. Bibliographies, abstracts, indices that list past and current publications in a field.
5. A writing manual for a particular field, if we know of a useful one.
6. A style manual that describes special features of citations, paper preparation, etc., if we know of a useful one.

Some books listed in (3), (5), and (6) may be out of print or available only through interlibrary loan. If there is no date listed in an item in (4), the publication appears annually. Sources marked NET are available in electronic form over the Internet. Sources marked CD are available on CD-ROMs with bibliographical data that in some cases go back several decades.

So rapid is technological change in the information sciences, however, that by the time you read this, new technology will have rendered much of our advice obsolete. Bibliographical sources that are not now in CD form or on-line will be. Or more likely, some new on-line resource will encompass all bibliographical sources. A new resource that will eventually revolutionize information gathering once again is the World Wide Web, or WWW. It will, in effect, create universes of connected information that you can roam through at will, if you know how. Your local bookstore will always have a new book to guide you through the resources of the Internet.

If you do not find what you are looking for in this list, keep in mind that for virtually every topic there is a dictionary and often an encyclopedia. So check in the catalogue under the general headings, *Dictionary of . . . , Encyclopedia of . . . ,* or *New Encyclopedia of . . . .*

GENERAL SOURCES

1. Blake, Lord, and C. S. Nicholls, eds. *The Dictionary of National Biography.* New York: Oxford University Press, 1990.
1. Chernow, Barbara A., and George A. Vallasi, eds. *The Columbia Encyclopedia.* 5th ed. New York: Columbia University Press, 1993.
1. Garraty, John A., ed. *Dictionary of American Biography.* New York: Scribner, 1994.
2. Goetz, Phillip W., ed. *The New Encyclopaedia Britannica.* 15th ed. 32 vols. Chicago: Encyclopaedia Britannica, 1987.
3. Kane, Eileen. *Doing Your Own Research: Basic Descriptive Research in the Social Sciences and Humanities.* New York: Marion Boyars Publishing, 1990.
3. Preece, Roy. *Starting Research: An Introduction to Academic Research and Dissertation Writing.* New York: St. Martin's Press, 1994.
3. Sheehy, E. P., ed. *Guide to Reference Work.* 10th ed. Chicago: American Library Association, 1986.
3. Vitale, Philip H. *Basic Tools of Research: An Annotated Guide for Students of English.* 3d ed., rev. and enl. New York: Barron's Educational Series, 1975.
4. *Biography Index: A Quarterly Index to Biographical Material in Books and Magazines.* New York: H. W. Wilson. (NET)
4. *Books in Print.* New York: R. R. Bowker. (NET, CD)
4. Charles, Dorothy, and Bea Joseph, eds. *Bibliographic Index: A Cumulative Bibliography of Bibliographies.* New York: H. W. Wilson. (NET)
4. *Dissertation Abstracts International.* Ann Arbor: UMI. (NET, CD)
4. *International Index.* New York: H. W. Wilson.
4. *Library of Congress Subject Catalog.* Washington, D.C.: Library of Congress. (NET, CD)
4. *National Newspaper Index.* Menlo Park, Calif.: Information Access. (NET, CD)
4. *New York Times Index.* New York: New York Times.
4. *Popular Periodical Index.* Camden, N.J.: Rutgers University.

4. *Readers' Guide to Periodical Literature.* New York: H. W. Wilson. (NET, CD)
4. *Subject Guide to Books in Print.* New York: R. R. Bowker. (NET, CD)
4. *Wall Street Journal Index.* New York: Dow Jones. (NET)
5. Williams, Joseph M. *Style: Toward Clarity and Grace.* Chicago: University of Chicago Press, 1995.
6. *The Chicago Manual of Style.* 14th ed. Chicago: University of Chicago Press, 1993.

SPECIAL SOURCES
Humanities

1. *Benet's Reader's Encyclopedia.* 3d ed. New York: Harper & Row, 1987.
3. *Humanities Index.* New York: H. W. Wilson. (NET, CD)
5. Barnett, Sylvan. *A Short Guide to Writing about Art.* 5th ed. New York: HarperCollins, 1996.

Art

1. Myers, Bernard S., ed. *McGraw-Hill Dictionary of Art.* New York: McGraw-Hill, 1969.
2. Myers, Bernard S., ed. *Encyclopedia of World Art.* New York: McGraw-Hill, 1983.
3. Arntzen, E., and R. Rainwater. *Guide to the Literature of Art History.* Chicago: American Library Association, 1980.
3. Jones, Lois Swan. *Art Information: Research Methods and Resources.* 3d ed. Dubuque: Kendall/Hunt, 1990.
3. Minor, Vernon Hyde. *Art History's History.* Englewood Cliffs: Prentice-Hall, 1994.
4. *Art Index.* New York: H. W. Wilson. (NET, CD)
5. Barnett, Sylvan. *A Short Guide to Writing about Art.* 5th ed. New York: HarperCollins, 1996.

History

1. Cook, Chris. *Macmillan Dictionary of Historical Terms.* 2d ed. London: Macmillan Reference, 1990.
1. Ritter, Harry. *Dictionary of Concepts in History.* Westport, Conn.: Greenwood Press, 1986.
2. Breisach, Ernst. *Historiography: Ancient, Medieval & Modern.* 2d ed. Chicago: University of Chicago Press, 1994.
3. Frick, E. *History: Illustrated Search Strategy and Sources.* 2d ed. Ann Arbor: Pierian, 1995.

3. Prucha, F. P. *Handbook for Research in American History.* 2d ed. Lincoln: University of Nebraska Press, 1994.
4. *Historial Abstracts.* Santa Barbara: ABC/CLIO. (NET, CD)
4. Kinnel, Susan., ed. *Historiography: An Annotated Bibliography of Journal Articles, Books, and Dissertations.* 2 vols. Santa Barbara: ABC-CLIO, 1987.
5. Barzun, J., and H. F. Graff. *The Modern Researcher.* 5th ed. New York: Harcourt Brace, 1992.

Literary Studies

1. Baldick, Chris. *Concise Oxford Dictionary of Literary Terms.* New York: Oxford University Press, 1991.
1. Brogan, T. V. F., ed. *The New Princeton Handbook of Poetic Terms.* Princeton: Princeton University Press, 1994.
1. Groden, Michael, and Martin Kreiswirth, eds. *The Johns Hopkins Guide to Literary Theory and Criticism.* Baltimore: The Johns Hopkins University Press, 1994.
1. Preminger, Alex, and T. V. F. Brogan, eds. *The New Princeton Encyclopedia of Poetry and Poetics.* Princeton: Princeton University Press, 1993.
2. Drabble, Margaret, ed. *Oxford Companion to English Literature.* Rev. ed. Oxford: Oxford University Press, 1995.
2. Hart, James David. *The Oxford Companion to American Literature.* 6th ed. New York: Oxford University Press, 1995.
3. Altick, R. A., and J. J. Fenstermaker. *The Art of Literary Research.* 4th ed. New York: Norton, 1993.
4. Blanck, Jacob. *Bibliography of American Literature.* New Haven: Yale University Press, 1991.
4. *Abstracts of English Studies.* Boulder, Colo.: National Council of Teachers of English.
4. *MLA International Bibliography of Books and Articles on the Modern Languages and Literature.* New York: MLA. (NET, CD)
5. Griffith, Kelly, Jr. *Writing Essays about Literature: A Guide and a Style Sheet.* 4th ed. Fort Worth: Harcourt Brace, 1994.
6. Gibaldi, Joseph. *MLA Handbook for Writers of Research Papers.* 4th ed. New York: MLA, 1995.

Music

1. Randel, Don Michael, ed. *The New Harvard Dictionary of Music.* Cambridge: Belknap Press of Harvard University Press, 1986.

1. Sadie, Stanley, ed. *The New Grove Dictionary of Music and Musicians.* 20 vols. New York: Macmillan, 1995.
2. Sadie, Stanley, ed. *The Norton Grove Concise Encyclopedia of Music.* Rev. and enl. New York: W. W. Norton, 1994.
3. Brockman, William S. *Music: A Guide To The Reference Literature.* Littleton, Colo.: Libraries Unlimited, 1987.
3. Duckles, Vincent H., and Michael A. Keller, eds. *Music Reference and Research Materials.* 4th ed., rev. New York: Schirmer, 1994.
4. *RILM Abstracts of Music Literature.* New York: RILM. (NET, CD)
4. *Music Index.* Detroit: Information Service. (NET, CD)
4. *Music Literature International.* National Information Services Co., NISC. (NET, CD)
5. Druesedow, John E. *Library Research Guide To Music: Illustrated Search Strategy And Sources.* Ann Arbor: Pierian Press, 1982.
5. Wingell, Richard J. *Writing about Music: An Introductory Guide.* Englewood Cliffs: Prentice-Hall, 1996.
6. Holomon, D. K. *Writing about Music: A Style Sheet from the Editors of 19th century Music.* Berkeley: University of California Press, 1988.

Philosophy

1. Blackburn, Simon. *The Oxford Dictionary of Philosophy.* New York: Oxford University Press, 1996.
1. Urmson, J. O., and Jonathan Ree, eds. *Concise Encyclopedia of Western Philosophy and Philosophers.* New ed. London: Routledge, 1993.
2. Parkinson, G. H. R. *The Handbook of Western Philosophy.* New York: Macmillan, 1988.
3. List, Charles, and Stephen H. Plum. *Library Research Guide to Philosophy.* Ann Arbor: Pierian Press, 1990.
4. *Dialog on Disc: Philosopher's Index.* Palo Alto, Calif.: DIALOG INFO. Services. (NET, CD)
4. *Philosopher's Index.* Bowling Green: Bowling Green University Press. (NET, CD)
5. Watson, R. *Writing Philosophy: A Guide to Professional Writing and Publishing.* Carbondale: Southern Illinois University Press, 1992.

Women's Studies

1. Bataille, Gretchen M., ed. *Native American Women: A Biographical Dictionary.* New York: Garland, 1993.

1. Mills, Jane. *Womanwords: A Dictionary of Words about Women.* New York: Holt, 1993.
1. Salem, Dorothy C., ed. *African American Women: A Biographical Dictionary.* New York: Garland, 1993.
1. Uglow, Jennifer S., and Frances Hinton, eds. *The Continuum Dictionary of Women's Biography.* New York: Continuum, 1989.
2. Hine, Darlene, et al., eds. *Black Women in America: An Historical Encyclopedia.* 2 vols. Brooklyn: Carlson, 1993.
2. Tierney, Helen, ed. *Women's Studies Encyclopedia.* 3 vols. New York: Greenwood Press, 1991.
2. Willard, Frances E., and Mary A. Livermore, eds. *American Women: Fifteen Hundred Biographies with Over 1,400 Portraits.* 2 vols. Rev. ed. Detroit: Gale Research, 1973.
3. Searing, Susan E. *Introduction to Library Research in Women's Studies.* Boulder: Westview Press, 1985.
4. *Women's Studies Abstracts.* Rush, N.Y.: Wilson.

Social Sciences

1. Sills, David, ed. *International Encyclopedia of the Social Sciences.* New York: Macmillan, 1991.
1. *Statistical Abstract of the United States.* Washington, D.C.: Bureau of the Census.
3. Oyen, Else, ed. *Comparative Methodology: Theory and Practice in International Social Research.* Newbury Park, Calif.: Sage, 1990.
3. Leith, Harr. *Bibliography For the Preparation of Research Papers in The History and Philosophy and Sociology of Science, Biography of Scientists, Science and Religion, Science and the Humanities, and Education in Science.* 7th ed. Toronto: York University, 1984.
3. Light, Richard J., and David B. Pillemer. *Summing Up: The Science of Reviewing Research.* Cambridge: Harvard University Press, 1984.
4. *Social Sciences Index.* New York: H. W. Wilson. (NET, CD)
5. Becker, Howard S. *Writing for Social Scientists: How to Start and Finish Your Thesis, Book, or Article.* Chicago: University of Chicago Press, 1986.
5. Bell, Judith. *Doing Your Research Project: A Guide for First-time Researchers in Education and Social Science.* 2d ed. Philadelphia: Open University Press, 1993.
5. Caryer, Ronald D. *Writing a Publishable Research Report in Education, Psychology, and Related Disciplines.* Springfield, Ill.: C. C. Thomas, 1984.

5. Krathwohl, David R. *How to Prepare a Research Proposal: Guidelines for Funding and Dissertations in the Social and Behavioral Sciences.* 3d ed. Syracuse: Syracuse University Press, 1988.

Anthropology

1. Winthrop, Robert H. *Dictionary of Concepts in Cultural Anthropology.* New York: Greenwood Press, 1991.
2. Ingold, Tim, ed. *Companion Encyclopedia of Anthropology.* New York: Routledge, 1994.
2. Levinson, David, ed. *Encyclopedia of World Cultures.* Boston: G. K. Hall, 1995.
3. Bernard, H. Russell. *Research Methods in Anthropology: Qualitative and Quantitative Approaches.* 2d ed. Thousand Oaks, Calif.: Sage Publications, 1994.
3. *Current Topics in Anthropology: Theory, Methods, and Content.* Reading: Addison-Wesley, 1972.
3. Glenn, James R. *Guide to the National Anthropological Archives.* Washington, D.C.: National Anthropological Archives, 1992.
4. *Abstracts in Anthropology.* Westport, Conn.: Greenwood Press.
4. *Annual Review of Anthropology.* Palo Alto: Annual Reviews.
5. Poggie, John J., Jr., et al. *Anthropological Research: Process and Application.* Albany: State University of New York Press, 1992.

Business

1. Link, Albert. *Link's International Dictionary of Business Economics.* Chicago: Probus Publications, 1993.
1. Nisburg, Jay N. *The Random House Dictionary of Business Terms.* New York: Random House, 1992.
1. Urdang, Laurence, ed. *Dictionary of Advertising: An Essential Resource for Professionals and Students.* Lincolnwood, Ill.: NTC Business Books, 1992.
2. Baker, W. H., ed. *Encyclopedia of American Business History and Biography.* 50 vols. New York: Facts on File, 1988–.
2. *Lifestyle Market Analyst: The Marketing Professional's Essential Source for Demographic and Lifestyle Activities.* Wilmette, Ill.: SRDS, 1995.
3. Cousin, Jill, and Lesley Robinson. *The Online Manual: A Practical Guide to Business Databases.* 3d ed. Oxford: Learned Information, 1994.
3. Daniells, L. M. *Business Information Sources.* 3d ed. Berkeley: University of California Press, 1993.

3. Woy, James B., ed. *Encyclopedia of Business Information Sources.* 10th ed. Detroit: Gale Research, 1996.
3. Kevin, John B. *Methods for Business Research.* New York: HarperCollins, 1992.
3. Sekaran, Uma. *Research Methods for Business.* 2d ed. New York: Wiley, 1992.
4. *Business Periodicals Index.* New York: H. W. Wilson. (NET, CD)
4. *United States Newspaper Program Database.* (NET)
5. Farrell, Thomas, and Charlotte Donabedian. *Writing the Business Research Paper.* Durham: Carolina Academic Press, 1991.
6. Vetter, William. *Business Law, Legal Research, and Writing: Handbook.* Needham Heights, Mass.: Ginn Press, 1991.

Communications and Journalism

1. Newton, Henry. *Newton's Telecom Dictionary: The Official Dictionary of Telecommunications, Networking and Voice Processing.* 9th ed. New York: Flatiron Pub., 1995.
1. Weik, Martin H. *Communications Standard Dictionary.* 3d ed. New York: Chapman Hall, 1996.
1. Weiner, Richard. *Webster's New World Dictionary of Media and Communications.* New York: Webster's New World, 1990.
2. Barnouw, E., ed. *International Encyclopedia of Communications.* 4 vols. New York: Oxford University Press, 1989.
2. Paneth, Donald. *The Encyclopedia of American Journalism.* New York: Facts on File, 1983.
2. Stern, Jane, and Michael Stern. *Encyclopedia of Pop Culture: An A to Z Guide of Who's Who and What's What, from Aerobics and Bubble Gum to Valley of the Dolls and Moon Unit Zappa.* New York: Harper Perennial, 1992.
3. Block, Eleanor, and James K. Bracken. *Communications and the Mass Media: A Guide to the Reference Literature.* Englewood, Colo.: Libraries Unlimited, 1991.
3. Blum, Eleanor, and Frances Goins Wilhoit. *Mass Media Bibliography. An Annotated Guide to Books and Journals for Research and Reference.* 3d ed. Urbana: University of Illinois Press, 1990.
3. Cates, S. A. *Journalism: A Guide to the Reference Literature.* Englewood, Colo.: Libraries Unlimited, 1990.
4. Lyle, Jack, et al., eds. *Communications Abstracts.* Los Angeles: University of California Press.

4. Matlon, Ronald J. *Index to Journals in Communication Studies Through 1990*. 2 vols. Annadale, Va.: Speech Communication Association, 1992.
6. Goldstein, Norm, ed. *The Associated Press Stylebook and Libel Manual: With Appendixes on Copyright Guidelines, Freedom of Information Act, Photo Captions, Filling the Wire*. 30th ed. NY: Associated Press, 1995.

Economics

1. Pearce, David W., ed. *MIT Dictionary of Modern Economics*. 4th ed. Cambridge: MIT Press, 1992.
2. Eatwell, John, et al., eds. *The New Palgrave: A Dictionary of Economics*. 4 vols. New York: Stockton, 1987.
2. Greenwald, Douglas, ed. *The McGraw-Hill Encyclopedia of Economics*. 2d ed. New York: McGraw-Hill, 1994.
3. Fletcher, J., ed. *Information Sources in Economics*. 2d ed. London: Butterworth, 1987.
3. Johnson, Glenn L. *Research Methodology for Economists: Philosophy and Practice*. New York: Macmillan, 1986.
4. *Journal of Economic Literature*. Nashville: American Economic Association. (NET, CD)
5. McCloskey, Donald. *The Writing of Economics*. New York: Macmillan, 1987.

Education

1. Lawton, Lewis. *Dictionary of Education*. Sevenoaks: Hodder & Stoughton, 1993.
1. Barrow, Robin. *A Critical Dictionary of Educational Concepts: An Appraisal of Selected Ideas and Issues in Educational Theory and Practice*. 2d ed. New York: Harvester Wheatsheaf, 1990.
2. Alkin, Marvin C., ed. *Encyclopedia of Educational Research*. 6th ed. 4 vols. New York: Macmillan, 1992.
2. Husen, T., and T. N. Postlewaite, eds. *International Encyclopedia of Education*. 2d ed. 12 vols. Tarrytown: Pergamon, 1994.
3. Buttlar, Lois. *Education: A Guide to Reference and Information Resources*. Englewood, Colo.: Libraries Unlimited, 1989.
3. *ERIC*. Resources in Education. (CD)
3. Keeves, John P., ed. *Educational Research, Methodology, and Measurement: An International Handbook*. New York: Pergamon Press, 1988.
3. Bausell, R. Barker. *Advanced Research Methodology: An Annotated Guide to Sources*. Metuchen, N.J.: Scarecrow Press, 1991.

4. *Current Index to Journals in Education.* New York: CCM, Information Sciences. (NET, CD)

4. *Education Index.* New York: H. W. Wilson, 1983. (NET, CD)

5. Tuckman, Bruce W. *Conducting Educational Research.* 4th ed. Fort Worth: Harcourt Brace College Publications, 1994.

6. Caryer, Ronald D. *Writing a Publishable Research Report in Education, Psychology, and Related Disciplines.* Springfield: C. C. Thomas, 1984.

Geography

1. Small, R. J. *A Modern Dictionary of Geography.* 3d ed. London: Edward Arnold, 1995.

2. Dunbar, Gary S. *Modern Geography: An Encyclopedic Survey.* New York: Garland, 1991.

2. Parker, Sybil P., ed. *World Geographical Encyclopedia.* New York: McGraw-Hill, 1995.

3. McKimmie, Timothy Irving. *How to Do Library Research in Geography.* Las Cruces: New Mexico State University Press, 1994.

3. Walford, Nigel. *Geographical Data Analysis.* New York: Wiley, 1995.

4. Okuno, Takashi. *A World Bibliography of Geographical Bibliographies.* Japan: Institute of Geoscience, 1992.

4. Conzen, Michael. *A Scholar's Guide to Geographical Writing on The American and Canadian Past.* Chicago: University of Chicago Press, 1993.

4. *Current Geographical Publications.* New York: American Geographical Society.

4. *Geographical Abstracts.* Norwich, England: University of East Anglia. (NET, CD)

5. Brandwein, Paul Franz. *The Earth: Research, Evaluation, and Writing.* New York: Harcourt Brace Jovanovich, 1980.

5. Durrenberger, Robert W. *Geographical Research and Writing.* New York: Cromwell, 1985.

6. Northey, Margot. *Making Sense in Geography and Environmental Studies: A Student's Guide to Research, Writing, and Style.* Toronto: Oxford University Press, 1992.

Law

1. Black, Henry C. *Black's Law Dictionary: Definitions of the Terms and Phrases of American and English Jurisprudence, Ancient and Modern.* 6th ed. St. Paul: West, 1991.

1. Curzun, L. B. *Dictionary of Law.* 4th ed. London: Pitman Publications, 1993.
1. *Guide to American Law.* 12 vols. and annual yearbooks. St. Paul: West, 1985.
2. *The Guide to American Law Supplement, 1995: Everyone's Legal Encyclopedia.* St. Paul: West, 1995.
2. Backer, Brian, and Patrick Petit, eds. *Encyclopedia of Legal Information Sources.* 2d ed. Detroit: Gale, 1993.
3. Amin, S. H. *Research Methods in Law.* Glasgow: Royston Publications, 1992.
3. Campbell, Enid, et al. *Legal Research: Materials and Methods.* 3d ed. North Ryde, N.S.W.: Law Book Co., 1988.
4. *Current Index to Legal Periodicals.* Seattle: M. G. Gallagher Law Library and Washington Law Review. (NET)
4. *Encyclopedia of Legal Information Sources: A Bibliographic Guide.* 2d ed. Detroit: Gale Research, 1993.
4. *Index to Legal Periodicals.* Bronx: H. W. Wilson, 1987. (CD)
4. *Lawdesk.* Rochester: Lawyers Cooperative Publications. (CD)
5. Bast, Carol M. *Legal Research and Writing.* Albany, N.Y.: Delmar Publishers, 1995.
6. *The Bluebook: A Uniform System of Citation.* 15th ed. Cambridge: Harvard Law Review, 1991.

Political Science

1. Robertson, David. *The Penguin Dictionary of Politics.* 2d ed. London: England: Penguin, 1993.
2. Barone, Michael, and Grant Ujifusa. *The Almanac of American Politics.* Washington, D.C.: National Journal.
2. Hawkesworth, M., and Maurice Kogan, eds. *Companion Encyclopedia of Government and Politics.* 2 vols. New York: Routledge, 1992.
2. Lal, Shiv, ed. *International Encyclopedia of Politics and Laws.* 17 vols. New Delhi: The Election Archives, 1987.
2. Miller, David, ed. *The Blackwell Encyclopaedia of Political Thought.* New York: Blackwell, 1987.
3. Baxter-Moore, Nicolas. *Studying Politics: An Introduction to Argument and Analysis.* Toronto: Copp Clark, Longman, 1994.
3. Holler, F. L., ed. *Information Sources of Political Science.* 4th ed. Santa Barbara: ABC/CLIO, 1986.

3. Johnson, Janet Buttolph. *Political Science Research Methods.* 3d ed. Washington, D.C.: Congressional Quarterly Press, 1995.
4. *ABC: Pol Sci.* Santa Barbara: ABC/Clio. (CD)
4. *PAIS International Journals Indexed.* New York: Public Affair Information Service. (NET, CD)
4. *U.S. Political Science Documents.* Pittsburgh: MidAtlantic Applications Center, University of Pittsburgh. (NET)
5. Lovell, David W. *Essay Writing and Style Guide for Politics and the Social Sciences.* Australia: Australian Political Studies Association, 1992.
6. Briddle, Arthur W. *Writer's Guide: Political Science.* Lexington, Mass.: D. C. Heath, 1987.

Psychology

1. Eysenck, Michael, ed. *The Blackwell Dictionary of Cognitive Psychology.* Oxford: Blackwell Preference, 1994.
1. Wolman, Benjamin, ed. and comp. *Dictionary of Behavioral Science.* 2d ed. San Diego: Academic Press, 1989.
1. Stratton, Peter. *A Student's Dictionary of Psychology,* 2d ed. London: E. Arnold, 1993.
2. Colman, Andrew M., ed. *Companion Encyclopedia of Psychology.* 2 vols. New York: Routledge, 1994.
2. Corsini, R. J., ed. *Encyclopedia of Psychology.* 2d ed. 4 vols. New York: Wiley, 1994.
3. Reed, J. G., and P. M. Baxter. *Library Use: A Handbook for Psychology.* 2d ed. Washington, D.C.: American Psychological Association, 1992.
3. Wilson, Christopher. *Research Methods in Psychology: An Introductory Laboratory Manual.* Dubuque: Kendall-Hunt, 1990.
4. *Annual Review of Psychology.* Palo Alto: Annual Reviews.
4. *Compact Cambridge MEDLINE.* Bethesda: NLM by Cambridge Scientific Abstracts. (CD)
4. *NASPSPA Abstracts.* Champaign: Human Kinetics Publishers.
4. *Psychological Abstracts.* Lancaster: American Psychological Association. (NET, CD)
5. Solomon, Paul R. *A Student's Guide to Research Report Writing in Psychology.* Glenview: Scott Foresman, 1985.
5. Sternberg, R. J. *The Psychologist's Companion: A Guide to Scientific Writing for Students and Researchers.* 3d ed. New York: Cambridge University Press, 1993.

6. *Publication Manual of the American Psychological Association.* 4th ed. Washington, D.C.: American Psychological Association, 1994.

Religion

1. Pye, Michael, ed. *Continuum Dictionary of Religion.* New York: Continuum, 1994.
1. Pye, Michael, ed. *Macmillan Dictionary of Religion.* London: Macmillan, 1994.
2. Eliade, M., ed. *Encyclopedia of Religion.* 16 vols. New York: Macmillan, 1987.
3. Kennedy, J. *Library Research Guide to Religion and Theology: Illustrated Search Strategy and Sources.* 2d ed., rev. Ann Arbor: Pierian, 1984.
4. Brown, David. *A Selective Bibliography of the Philosophy of Religion.* Oxford: Sub-Faculty of Philosophy, 1990.
4. Chinyamu, Salms. *An Annotated Bibliography on Religion.* Malawi: Malawi Library Association, 1993.
4. *ATLA Religion Database on CD-ROM.* Evanston: American Theological Library Association.
4. *Religion Index One/Two: Periodicals, Religion, and Theological Abstracts.* Chicago: American Theological Library Association. (NET, CD)
4. *Religion Studies: Bibliography of Material Sources for the Course.* North Sydney: Board of Studies, NSW, 1990.

Sociology

1. Abercrombie, Nicholas. *The Penguin Dictionary of Sociology.* 3d ed. London: Penguin Books, 1994.
1. Gordon, Marshall, ed. *The Concise Oxford Dictionary of Sociology.* Oxford: Oxford University Press, 1994.
2. Borgatta, Edgar F., ed. *Encyclopedia of Sociology.* 4 vols. New York: Macmillan, 1992.
2. Smelser, N., ed. *Handbook of Sociology.* Newbury Park, Calif.: Sage, 1988.
3. Aby, S., ed. *Sociology: A Guide to Reference and Information Sources.* Englewood, Colo.: Libraries Unlimited, 1987.
3. Oyen, Else, ed. *Comparative Methodology: Theory and Practice in International Social Research.* Newbury Park, Calif.: Sage, 1990.
4. *Annual Review of Sociology.* Palo Alto: Annual Reviews.
4. *ASSIA.* London: Bowker Saur. (NET, CD)

4. *Social Science Research.* Philadelphia: Institute for Scientific Information.
4. *Social Sciences Index.* New York: H. W. Wilson. (NET, CD)
4. *Sociological Abstracts.* New York: Sociological Abstracts. (NET, CD)
5. Tomovic, V., ed. *Definitions in Sociology: Convergence, Conflict, and Alternative Vocabularies: A Manual for Writers of Term Papers, Research Reports, and Theses.* St. Catharine's, Ont.: Diliton Publications, 1979.
5. Giarrusso, Roseann, et al. *A Guide to Writing Sociology Papers,* ed. Judith Richlin-Klonsky and Ellen Strenski. 3d ed. New York: St. Martin's Press, 1994.

Natural Sciences

1. *McGraw-Hill Science And Technical Reference Set.* Release 2.0. New York: McGraw-Hill, 1992. (CD)
1. Morris, Christopher, ed. *Academic Press Dictionary of Science and Technology.* San Diego: Academic Press, 1992.
1. Walker, Peter M. B., ed. *Chambers Science and Technology Dictionary.* Edinburgh: Chambers, 1991.
2. *McGraw-Hill Multimedia Encyclopedia of Science and Technology.* New York: McGraw-Hill, 1994. (CD)
3. *Directory of Technical and Scientific Directories: A World Bibliographic Guide to Medical, Agricultural, Industrial, and Natural Science Directories.* 6th ed. Phoenix: Oryx Press, 1989.
3. Nielsen, Harry A. *Methods of Natural Science: An Introduction.* Englewood Cliffs: Prentice-Hall, 1967.
4. *Science Citation Index with Abstracts.* Philadelphia: Institute for Scientific Information. (NET, CD)
4. *Wilson General Science Abstracts.* Bronx: H. W. Wilson. (NET, CD)
5. Booth, Vernon. *Communicating in Science: Writing a Scientific Paper and Speaking at Scientific Meetings.* 2d ed. Cambridge: Cambridge University Press, 1993.
5. Woodford, F. Peter, ed. *Scientific Writing for Graduate Students: A Manual on the Teaching of Scientific Writing.* Committee on Graduate Training in Scientific Writing. Bethesda: Council of Biology Editors, 1986.
5. Style Manual Comm., Council of Biology Editors. *Scientific Style and Format: The CBE Manual for Authors, Editors, and Publishers.* 6th ed. Cambridge: Cambridge University Press, 1994.

Biology

1. *A Concise Dictionary of Biology.* New ed. Oxford: Oxford University Press, 1990.
1. Allaby, Michael, ed. *The Oxford Dictionary of Natural History.* Oxford: Oxford University Press, 1985.
1. Singleton, Paul, and Diana Sainsbury. *Dictionary Of Microbiology and Molecular Biology.* 2d ed. New York: Wiley, 1993.
2. *Biology Encyclopedia.* New York: HarperCollins, 1991. (Videodisc)
2. Dulbecco, Renato, ed. *Encyclopedia of Human Biology.* San Diego: Academic Press, 1991.
2. Kendrew, John, ed. *Encyclopedia of Molecular Biology.* Cambridge, Mass.: Blackwell Science, 1995.
3. Roper, Fred W., and Jo Anne Boorkman. *Introduction to Reference Sources in Health Sciences.* 3d ed. Chicago: Medical Library Association, 1994.
3. Wyatt, H. V., ed. *Information Sources in the Life Sciences.* 4th ed. London: Bowker-Saur, 1994.
4. *Biological Abstracts on Compact Disc CD-Rom.* Philadelphia: Biological Abstracts. (NET, CD)
4. *Biological and Agricultural Index.* Bronx: H. W. Wilson. Updated quarterly. (NET, CD)
4. *Environmental Abstracts.* Bethesda: Congressional Information Service. (NET, CD)
5. Brooks, William Stewart. *Writing in The Biological Sciences.* Ripon, Wisc.: Department of Biology, Ripon College, 1990.
6. *CBE Style Manual: A Guide for Authors, Editors, and Publishers in the Biological Sciences.* 5th ed. Bethesda: Council of Biology Editors, 1983.

Chemistry

1. *CRC Handbook of Chemistry and Physics.* 76th ed. Boca Raton: CRC Press, 1995.
1. Wenske, Gerhard. *Dictionary of Chemistry: German/English.* New York: VCH, 1994.
2. Kroschwitz, Jacqueline I., and Mary Howe-Grant, eds. *Kirk-Othmer Encyclopedia of Chemical Technology.* 4th ed. New York: Wiley, 1993.
2. Meyers, Robert A., ed. *Encyclopedia of Physical Science and Technology.* 2d ed. San Diego: Academic, 1992.
3. Leslie, Davies. *Efficiency in Research, Development, and Production: The Sta-*

tistical Design and Analysis of Chemical Experiments. Cambridge: Royal Society of Chemistry, 1993.

3. Wiggins, Gary. *Chemical Information Sources*. New York: McGraw-Hill, 1991.

4. *Chemical Abstracts*. Washington, D.C.: American Chemical Society. (NET)

4. *Composite Index for CRC Handbooks*. 3d ed. Boca Raton: CRC Press, 1991. (CD)

5. Dodd, Janet S., ed. *The ACS Style Guide: A Manual for Authors and Editors*. Washington, D.C.: American Chemical Society, 1986.

6. Schoenfeld, Robert. *The Chemist's English, With "Say It in English, Please!"* 3d rev. ed. New York: VCH, 1989.

Computer Sciences

1. South, David W. *The Computer and Information Science and Technology Abbreviations and Acronyms Dictionary*. Boca Raton: CRC Press, 1994.

1. Spencer, Donald. *Webster's New World Dictionary of Computer Terms*. 5th ed. New York: Macmillan, 1994.

2. Ralston, Anthony, and Edwin D. Reilly. *Encyclopedia of Computer Science*. 3d ed. New York: International Thomson Computer Press, 1995.

3. Ardis, Susan B. *A Guide to the Literature of Electrical and Electronics Engineering*. Ed. Jean M. Poland. Littleton, Colo.: Libraries Unlimited, 1987.

4. *Applied Science and Technology Index*. New York: H. W. Wilson. (NET, CD)

4. Cibbarelli, Pamela R., ed. *Directory of Library Automation Software, Systems, and Services*. Medford, N.J.: Learned Information, 1994.

5. Eckstein, C. J. *Style Manual For Use in Computer-Based Instruction*. Brooks Air Force Base, Texas: Air Force Human Resources Laboratory, Air Force Systems Command, 1990.

Geology

1. Bates, Robert Latimer, and Julia A. Jackson, eds. *Glossary of Geology*. 3d ed. Alexandria, Va.: American Geological Institute, 1987.

1. Challinor, John. *Challinor's Dictionary of Geology*. 6th ed. Ed. Anthony Wyatt. Cardiff: University of Wales Press, 1986.

2. Nierenberg, William A., ed. *Encyclopedia of Earth System Science*. 4 vols. San Diego: Academic Press, 1992.

2. Parker, Sybil P., ed. *McGraw-Hill Encyclopedia of the Geological Sciences.* 2d ed. New York: McGraw-Hill, 1988.

2. Smith, David G., ed. *The Cambridge Encyclopedia of the Earth Sciences.* Cambridge: Cambridge University Press, 1982.

3. Wood, David Norris, ed. *Use of Earth Sciences Literature.* London: Butterworth, 1973.

4. *Bibliography and Index of Geology.* Falls Church, Va.: American Geological Institute. (NET, CD)

4. *Geo Abstracts.* Norwich: Geo Abstracts. (NET, CD)

4. *Geobase.* Norwood, Mass.: Silver Platter. (CD)

5. Dunn, J., et al. *Organization and Content of a Typical Geologic Report.* Rev. ed. Arvada, Colo.: American Institute of Professional Gerologists, 1993.

6. Cochran, Wendell et al. *Geowriting: A Guide to Writing, Editing, and Printing in Earth Science.* 4th ed. Alexandria, Va.: American Geological Institute, 1984.

Mathematics

1. Borowski, E. J., J. M. Borwein, et al., eds. *HarperCollins Dictionary of Mathematics.* New York: Harper Perennial Books, 1991.

1. James, Robert Clarke. *Mathematics Dictionary.* 5th ed. New York: Van Nostrand Reinhold, 1992.

1. Schwartzman, Steven. *The Words of Mathematics: An Etymological Dictionary of Mathematical Terms Used in English.* Washington, D.C.: Mathematical Association of America, 1994.

2. Ito, Kiyosi, ed. *Encyclopedic Dictionary of Mathematics.* 2d ed. 2 vols. Cambridge: MIT Press, 1993.

3. Pemberton, John E. *How to Find Out in Mathematics.* 2d rev. ed. Oxford: Pergamon Press, 1969.

4. *East European Scientific Abstracts.* Arlington, Va.: JPRS.

4. *Mathematical Reviews: 50th Anniversary Celebration,* Providence, R.I.: American Mathematical Society, 1990.

4. *Mathsci.* Providence, R.I.: American Mathematical Society. (NET, CD)

5. *A Manual For Authors Of Mathematical Papers.* Rev. ed. Providence, R.I.: American Mathematical Society, 1990.

Physics

1. Sube, Ralf. *Dictionary, Physics Basic Terms: English-German.* Berlin: A. Hatier, 1994.

1. Thewlis, James. *Concise Dictionary of Physics and Related Subjects.* 2d ed. rev. and enl. Oxford: Pergamon Press, 1979.

2. Lerner, Rita G., and George L. Trigg, eds. *The Encyclopedia of Physics.* 2d ed. New York: VCH, 1991.

2. Parker, Sybil P., ed. *McGraw-Hill Encyclopedia of Physics.* 2d ed. New York: McGraw-Hill, 1993.

2. Meyers, Robert A., ed. *Encyclopedia of Modern Physics.* San Diego: Academic Press, 1990.

2. Trigg, George L., et al., eds. *Encyclopedia of Applied Physics.* New York: VCH Publishers, 1996.

3. Shaw, Dennis F. *Information Sources in Physics.* 3d ed. New Jersey: Bowker-Saur, 1994.

3. Malinowsky, Robert H. *Science and Technology Information Sourcebook.* Phoenix: Oryx Press, 1994.

4. *Applied Science and Technology Index.* New York: H. W. Wilson. (CD)

4. Bohme, Siegfried. *Astronomy and Astrophysics Abstracts.* Berlin: Springer-Verlag, 1985.

4. *Current Physics Index.* Westbury, N.Y.: American Institute of Physics. (NET, CD)

4. *Physics Abstracts.* London: Institute of Electrical Engineers.

5. Katz, M. J. *Elements of the Scientific Paper: A Step-by-Step Guide for Students and Professionals.* New Haven: Yale University Press, 1985.

6. American Institute of Physics. *AIP Style Manual.* 4th ed. Westbury, N.Y.: American Institute of Physics, 1990.

Index